American put options

Pitman Monographs and
Surveys in Pure and Applied Mathematics 84

American put options

D M Salopek

The Fields Institute

LONGMAN

Addison Wesley Longman Limited
Edinburgh Gate, Harlow
Essex CM20 2JE, England
and Associated companies throughout the world.

Published in the United States of America
by Addison Wesley Longman Inc.

First published 1997

AMS Subject Classifications: (Main) 90-02, 90A09
(Subsidiary) 60H30

ISSN 0269-3666

ISBN 0 582 31594 8

British Library Cataloguing in Publication Data

A catalogue record for this book is
available from the British Library

Printed and bound by Bookcraft (Bath) Ltd

To my parents,

Frank and Mary

Table of Contents

Preface

The purpose of this book to provide a detailed overview of recent works on the American put option from both theoretical and computational approaches. It can be used as an introduction to the American put for both mathematicians and finanical students. This book assumes the reader has some knowledge of martingale theory, stochastic analysis, finance, partial differential equations and stochastic partial differential equations.

In Chapter 1, we briefly describe Bachelier's model and go into detail of the Black-Scholes model. The we derive the differential equations characterizing the American put, and its generalizations. In Chapter 2, Jacka's results on optimal stopping and the free boundary formulation are presented. In Chapter 3 and Chapter 4, several analytical approximations and numerical methods, respectively, to the American put value are also given and analyzed.

The first version of this book was written in 1994 as a part of a Masters degree under the supervision of Robert J Elliott. There has been several new developments in the numerical analysis of the American put since then. We have provided an additional bibliography at the end of Chapter 4 so the reader can be familiar with these works as well.

There are many who helped me prepare this book, either through discussions or by reading of the manuscript. I would like to thank Donald Dawson, Thomas Mikosch, Byron Schmuland, John van der Hoek, and Augustine Wong. I would especially like to thank Robert Elliott for introducing me to this topic and mathematical finance. I am also extremely grateful to both Rimas Norvaiša and Mazi Shirvani for their criticisms and proofreading of the manuscript. As for the mistakes and misprints, they are all mine.

I am greatly indebted to both Sarah Dahabieh and Donald Dawson for providing me an excellent working environment to revise this book at Carleton University.

December 1996 *D M Salopek*

Chapter 1
Introduction

The theory of option pricing began in 1900 when Bachelier [2] deduced an option pricing formula based on the assumption that stock prices follow what we now call standard Brownian motion. This nearly forgotten work was Bachelier's doctoral dissertation, *Theory of Speculation.* In this work, Bachelier describes a random walk model of stock price change which we will discuss below.

Bachelier assumed that the stock price at time t is a random variable, S_t, and these price changes over equal and non-overlapping time intervals are independent and identically distributed such that

$$\Pr(S_{t+T} \leq u \mid S_t = x) = F_T(u - x) \tag{1.1}$$

for all $u, x \in \mathbf{R}$, and all $t, T \geq 0$, where F_T is a distribution function. However, Bachelier incorrectly deduced that (1.1) implies the following,

$$F_T(u - x) = \Phi\left(\frac{u - (x + \mu T)}{\sigma\sqrt{T}}\right), \tag{1.2}$$

where μ is the mean price change per time period, σ^2 is the variance per time period, and Φ denotes the distribution function of a standard normal random variable. It is not sufficient to deduce (1.2) from (1.1) since it is known that a distribution function of any stable non-gaussian random variable[1] (see [18]) also satisfies (1.1). Equation (1.2) follows from (1.1) if the variance of each S_t is finite.

Furthermore, Bachelier assumed that the mean price change per unit time μ is zero, since he viewed stock market transactions as a

[1] In the finance literature, this distribution is often called the stable Paretian distribution.

gamble, and felt that competition would reduce the expected return to zero. This is unsuitable as an equilibrium specification, since it seems to deny both positive interest rates and the practice of risk aversion.

In addition, because $\mu = 0$, $\Pr(S_T < u)$ tends to 1/2 for every u as T tends to infinity. Indeed, let $t = 0$. Then, by the law of total probability, we have, for every $u \in \mathbf{R}$,

$$\begin{aligned}
\Pr(S_T \leq u) &= \int_{-\infty}^{\infty} \Pr(S_T \leq u \mid S_0 = x)\, dG(x) \\
\text{by (1.1) and (1.2)} &= \int_{-\infty}^{\infty} \Phi\left(\frac{u-x}{\sigma\sqrt{T}}\right) dG(x) \\
&\overset{T\to\infty}{\longrightarrow} \Phi(0) \int_{-\infty}^{\infty} dG(x) = \frac{1}{2},
\end{aligned}$$

where G is the distribution function of S_0. Since nothing in this calculation restricts u to be a positive number, there is a positive probability of negative stock prices. This is a violation of the property of limited liability, where *limited liability* states that the investor cannot lose more than the amount he has paid for his shares. Therefore, prices can never fall below zero.

Now we apply Bachelier's assumptions to calculate a *call option*, which is the right to buy one share at time T with *exercise price* X. We show that his formula lead to unrealistic outcomes. Assume that the call option value at time r is the function $c_r \equiv c(S_r, T - r; X)$, where $S = \{S_r : 0 \leq r \leq T\}$ is the price process satisfying (1.1). Then the *fair price* of the call option is assumed to be $c^* \equiv E[c(S_T, 0; X)]$, where

$$c(S_T, 0; X) = (S_T - X)^+ = \max(S_T - X, 0). \tag{1.3}$$

Let S_0 be a positive constant. Since $\mu = 0$, we have

$$c^* = \int_X^{\infty} (u - X)\, d\Phi \left(\frac{u - S_0}{\sigma\sqrt{T}} \right). \tag{1.4}$$

Indeed, by the law of total probability for expectations, we get

$$\begin{aligned}
E(S_T - X)^+ &= \int_{-\infty}^{\infty} E\{(S_T - X)^+ \mid S_0 = x)\}\, dG_0(x) \\
\text{by (1.1) and (1.2)} &= \int_{-\infty}^{\infty} \int_X^{\infty} (u - X)\, d\Phi \left(\frac{u - x}{\sigma\sqrt{T}} \right) dG_0(x) \\
&= \int_X^{\infty} (u - X)\, d\Phi \left(\frac{u - S_0}{\sigma\sqrt{T}} \right),
\end{aligned}$$

where

$$G_0(x) = \begin{cases} 0 & \text{if } x < S_0 \\ 1 & \text{if } x \geq S_0. \end{cases}$$

Therefore (1.4) holds. By the change of variables $v \equiv (u - S_0)/\sigma\sqrt{T}$, we get

$$c^* = \int_{(X-S_0)/\sigma\sqrt{T}}^{\infty} (v\sigma\sqrt{T} + S_0 - X)\, d\Phi(v). \tag{1.5}$$

To evaluate this integral, first note that

$$\begin{aligned}
\Phi''(x) &= [\Phi'(x)]' = [\frac{1}{\sqrt{2\pi}} e^{-\frac{x^2}{2}}]' \\
&= \frac{1}{\sqrt{2\pi}} e^{-\frac{x^2}{2}} (-x) = -x\Phi'(x)
\end{aligned} \tag{1.6}$$

for all $x \in \mathbf{R}$. Thus, for every $a \in \mathbf{R}$,

$$\begin{aligned}
\int_a^{\infty} v d\Phi(v) &= \int_a^{\infty} v\Phi'(v) dv \\
\text{by (1.6)} &= -\int_a^{\infty} \Phi''(v) dv \\
&= -\int_a^{\infty} d\Phi'(v) \\
&= \Phi'(a) = \Phi'(-a).
\end{aligned}$$

This allows us to write (1.5) as follows

$$c^* = S_0 \cdot \Phi\left(\frac{S_0 - X}{\sigma\sqrt{T}}\right) - X \cdot \Phi\left(\frac{S_0 - X}{\sigma\sqrt{T}}\right) + \sigma\sqrt{T}\Phi'\left(\frac{S_0 - X}{\sigma\sqrt{T}}\right). \quad (1.7)$$

On further examination of (1.7), we observe that as time tends to infinity, the call price increases without bound. So, (1.7) does not satisfy a reasonable condition that the maximum value of a call option should not exceed the stock price. This is because of the lack of the limited liability assumption.

In summary, the drawbacks of Bachelier's model are

1. The assumption that price movements follow standard Brownian motion gives a nonzero probability of negative prices for the stock, and the possibility that option prices may be greater than their respective stock prices for large T.

2. The assumption that the mean price changes μ is zero ignores the investor's preferences to be risk neutral.

3. The implicit assumption that the variance is finite is only applicable to the normal distribution, and so it excludes other stable non-gaussian distributions.

Despite these flaws, the Bachelier's model provides great insight into the study of options. In his work, one sees unexpectedly that Bachelier has discovered the notion of an efficient market, a concept that popularized in the 1960's. For an historical perspective, Bachelier's dissertation is still worth reading.

It should be noted that the call option which was defined by (1.3) is known as a European option (see definition below) under the assumption that stock price changes follow standard Brownian motion.

In the coming sections, we will present the modern perspective of option pricing. In particular, we will review the celebrated work of Black and Scholes [5]. In their model, they assume that stock price changes follow a geometric Brownian motion which prevents the stock prices from falling below zero.

1.1 Why study options?

An option is a particularly simple type of the contingent claim asset, where a *contingent claim* is a contract or agreement which pays a specified amount at time T if a certain event happens. While options are very specialized financial instruments, the same analysis could be applied to the pricing of a firm's liabilities, the term and risk structure of interest rates, and the theory of speculative markets (see Merton [46] for further discussion on this). An additional incentive to study option pricing theory is the possibility to test one's theories from empirical data. Before we proceed further, we will define the specific types of options.

DEFINITION. An *American warrant* is a security, issued by a company, giving its owner the right to purchase a share of stock at a given exercise price on or before a given date. An *American call option* has the same terms as the warrant except that it is issued by an individual instead of a company. An *American put option* gives its owner the right to sell a share of stock at a given exercise price on or before a given date. A *European option* has the same terms as its American counterpart except that it cannot be exercised before the last date of the contract.

Samuelson [50] showed that the American and European options may not have the same value. The contracts may differ with respect to other provisions such as antidilution clauses, exercise price changes, etc. We have attempted to collect a survey of the what has been done on the American put option.

In this section, we assume that the $S = \{S_t : 0 \leq t \leq T\}$ denotes a stock price process and it does not have a specific distribution unless otherwise stated. We would like to note that any assumptions about the distribution of the stock price are essential in mathematical finance. Under the modern theory, it is assumed that the stock price is a semimartingale (see Karatzas [31] for further details). However, in the subsequent sections, we will restrict ourselves by assuming the stock price process to be geometric Brownian motion, which is a simple example of semimartingales.

The following notation will be used throughout the survey:

- t = current time and $0 \leq t \leq T$.
- T = expiration date of an option.
- $T - t$ = the time to expiration.
- r = risk-free interest rate.
- X = exercise (strike) price of an option.
- S_t = stock price at time t.
- S_c = the critical stock price = the price at which the put option should be exercised.
- P_t = price of the American put at time t.
- C_t = price of the American call at time t.
- p_t = price of the European put at time t.
- c_t = price of the European call at time t.

Assumption 1 *We will assume that all options are functions of stock price, the time to expiration and the exercise price. That is,* $P_t = P(S_t, T-t; X)$, $p_t = p(S_t, T-t; X)$, $C_t = C(S_t, T-t; X)$, *and* $c_t = c(S_t, T-t; X)$, *for* $0 \leq t \leq T$.

Merton [44] deduces a set of restrictions on option pricing formulae from the assumption that investors prefer more to less (a necessary condition for the formula to be consistent with a pricing theory). However, these restrictions are not sufficient to uniquely determine an option price formula, so additional assumptions are required. In this section, we will review these assumptions and definitions to formulate the problem of option pricing. We will start by giving several definitions.

DEFINITION 1. Security A is *dominant* over security B if the return on A is never less than the return on B, and is greater than that on B at some known date in the future (see [44]).

DEFINITION 2. *Short selling* refers to an opportunity to sell a security even if one does not own the security. That is, one borrows the security from an investor who has the security and is obligated to buy back the equal number of shares from the lender.

DEFINITION 3. An *Arbitrage opportunity* is a riskless plan to make profits without investment.

In the case of *ideal* markets with no transaction costs and the ability to borrow and short sell without restriction, the existence of dominated securities would be equivalent to the existence of arbitrage opportunities. Alternatively for less ideal markets, it is possible to have dominated securities without arbitrage opportunities. In this case, we need to assume "symmetric market rationality" (see [47], p. 427) and that investors prefer more wealth to less (see Merton [46], p. 257).

Assumption 2 *The option to be priced is neither a dominant nor a dominated security.*

This is a necessary condition for an option pricing theory without arbitrage.

THE EUROPEAN PUT OPTION PRICE

At first, the put option received relatively little analysis in the financial literature because it is a less popular option than the call and, because it is generally assumed (see Black and Scholes [5]) that given the price of a call option and the underlying stock, the value of a put is uniquely determined. We now know this is false for American put options (see, for example, Merton [46]) . In particular, the mathematics of put option pricing is more complicated than that of the corresponding call option. However, for the European option this is true and will be illustrated next. This section primarily follows Merton [44].

Let $B(T)$ be the price of a riskless, discounted loan (or bond), which pays \$1, T years from now, and $B'(T)$ the current value of \$1 payable T years from now at the borrowing rate of a T-year, noncallable, discounted loan. Since it is assumed that current and future interest rates are positive, it follows that $1 = B(0) > B(t_1) > B(t_2) > \ldots > B(t_n)$ whenever $0 < t_1 < t_2 < \ldots < t_n$. To avoid arbitrage, $B'(T) \leq B(T)$. Recall $T - t$ is the time to expiration.

At the expiration date of a put option, we have

$$P(S_T, 0; X) = p(S_T, 0; X) = (X - S_T)^+. \tag{1.8}$$

Consider the following two portfolio strategies to price the European put option. In the first, we take a long position (to buy) in the underlying stock at S_t, a long position in a T-year European put at $p(S_t, T-t; X)$, and borrow $XB'(T)$. The value of this portfolio at time

T will be: $S_T + (X - S_T) - X = 0$, if $S_T \leq X$, and $S_T + 0 - X = S_T - X$ if $S_T > X$. The pay-off structure is identical in every respect to a European call option with the same exercise price and duration. Therefore, to stop the call option from being a dominated security, the put and the call must be priced so that

$$p(S_t, T-t; X) + S - XB'(T-t) \geq c(S_t, T-t; X). \qquad (1.9)$$

For the second portfolio strategy, we take a long position (to buy) in a T-year European call, a short position (to sell) in the underlying stock price S_t, and lend $XB(T)$ dollars. The value of this portfolio at time T will be: $0 - S_T + X = X - S_T$, if $S_T \leq X$, and $(S_T - X) - S_T + X = 0$ if $S_T > X$. Similarly, this pay-off structure is identical in every respect to a European put option with the same exercise price and duration. To prevent the put from being a dominated security, we also require

$$c(S_t, T-t; X) - S_t + XB(T-t) \geq p(S_t, T-t; X) \qquad (1.10)$$

must hold. We now can establish the put-call parity.

Theorem 1 *If Assumption 2 holds and if the borrowing and lending rates are equal (i.e., $B(T-t) = B'(T-t)$ for all $t, T \geq 0$), then*

$$p(S_t, T-t; X) = c(S_t, T-t; X) - S_t + XB(T-t). \qquad (1.11)$$

PROOF. Since $B(T) = B'(T)$, apply simultaneously (1.9) and (1.10) to establish equality. □

From the above theorem, we can determine the price of an European put option once we have the value of European call option. Notice that in the proof of Theorem 1, we did not require any distributional assumptions on the stock price changes or future interest rates. With the above analysis, we have the following two corollaries to Theorem 1.

Corollary 1 $p(S_t, T-t; X) \leq XB(T-t)$.

PROOF. Since $S_t \geq C(S_t, T-t; X) \geq c(S_t, T-t; X)$, we have $c_t - S_t \leq 0$, so the result follows from (1.10). □

Corollary 2 *The value of a perpetual (i.e., $T = \infty$) European put option is zero.*

PROOF. Recall that a put is a limited liability security (i.e., $p(S_t, T-t; X) \geq 0$). From Corollary 1 and the condition that $B(\infty) = 0$, one obtains $0 \geq p(S_t, \infty; X)$. □

This is appropriate for the European put option analysis, however, in the case of the American put option, we have the right to exercise the option at any time before the expiration time. Hence, its price must satisfy the arbitrage condition

$$P_t \geq (X - S_t)^+. \tag{1.12}$$

In addition, it also follows from Assumption 2 that

$$P(S_t, T-t; X) \geq p(S_t, T-t; X), \tag{1.13}$$

where the strict inequality holds only if there is a positive probability of exercising early. Hence, unlike a European option, the value of an American option is always a nondecreasing function of its expiration date. If there is no possibility of exercising prematurely, then the value of the American option will be equal to its European counterpart. Also, if the strike price is constant and no dividends are paid, then the European and American call options will have the same value. Even under these assumptions, there is usually a positive probability of prematurely exercising an American put, so that the American put will sell for more than its European counterpart (see Merton [44]). For instance, the right to the early exercise of an American put becomes more attractive as the current stock price increases. Therefore, we have the following theorem:

Theorem 2 *If, for some $\tau < T$, there is a positive probability that $c(S_0, \tau; X) < X(1 - B(\tau))$, then there is a positive probability that an American put option with expiration time T will be exercised prematurely, and the value of the American put will strictly exceed the value of its European counterpart.*

PROOF. The only reason that an American put will sell at a higher price than its European counterpart is because there is a positive probability of exercising prior to expiration. Hence, it is sufficient to prove that $p(S_t, T-t; X) < P(S_t, T-t; X)$. By Assumption 2, if for some $\tau \le T$, $p(S_0, \tau; X) < P(S_0, \tau; X)$ for some value of S_0, then $p(S_t, T-t; X) < P(S_t, T-t; X)$. By assumption, there exists S_0 such that $c(S_0, \tau; X) < X(1 - B(\tau))$. Then Theorem 1 implies that $p(S_0, \tau; X) = c(S_0, \tau; X) - S_0 + XB(\tau) < X - XB(\tau) - S_0 + XB(\tau) = X - S_0 \le (X - S_0)^+ \le P(S_0, \tau; X)$. □

Since there is almost always a chance of exercising early, the formula in Theorem 1 will not lead to a correct valuation of an American put. Furthermore, the valuation of an American put option is analytically more difficult than its European counterpart, as we will see from this survey.

1.2 The Black-Scholes Model

A number of option pricing theories satisfy the modern approach to option pricing. One such theory developed by Black and Scholes [5] is particularly attractive because it is a complete general equilibrium formulation of the problem and because the final formula is a function of observable variables, making the model subject to direct empirical tests.

The assumptions of Black-Scholes model are as follows:

1. The risk-free interest rate, r, is known and constant through time.

2. The stock price $S = \{S_t : 0 \leq t \leq T\}$ is a geometric Brownian motion. That is,

$$S_t = S_0 e^{\sigma W_t + (\mu - \sigma^2/2)t}, \tag{1.14}$$

   where S_0 denotes the stock price process at time 0, $W = \{W_t : 0 \leq t \leq T\}$ denotes Brownian motion, μ denotes the expected return of the stock price changes, and σ denotes the volatility (standard deviation) of stock price changes.

3. No dividends are paid, and the exercise price does not change over the life of the contract.

4. The option is European so that it can only be exercised at maturity.

5. There are no transaction costs in buying or selling the stock or the option and there are no taxes.

6. It is possible to borrow any fraction of the price of a security to buy it or to hold it, at the risk-free interest rate.

7. There are no penalties to short selling. A seller who does not own a security will simply accept the price of the security from a buyer, and will agree to settle with the buyer at some future date by paying her an amount equal to the price of the security on that date.

REMARK: Samuelson in the 1960's (see [50]) first introduced geometric Brownian as a possible representation of the stock price changes.

In the finance literature, geometric Brownian motion is often called a lognormal distribution.

Black and Scholes used Samuelson's theory of warrant pricing [50] to represent the expected return on the option in terms of the option price function and its partial derivatives. This was possible since they assumed that an option price is a function of the underlying stock price and time to expiration. Their strategy was to reduce the portfolio's exposure to market risks (i.e., hedge the portfolio). They did this by choosing appropriate quantities of the underlying stock, the option, and a short-term riskless bond. Now according to the Capital Asset Pricing Model (CAPM, see Sharpe [53]), any portfolio with zero market risk must give $\mu = r$, that is the expected return of the stock is equal to riskless interest rate. Hence, there exists an equilibrium condition between the expected return of the stock μ and the expected return of the option. Therefore, from this equilibrium condition on option yield, a partial differential equation can be derived for the option price (see (1.25)). The Black-Scholes' solution for the European call option is

$$c(S_t, T-t; X) = S_t\Phi(d_1) - Xe^{-r(T-t)}\Phi(d_2), \tag{1.15}$$

where

$$d_1 = [\log(S_t/X) + (r+\sigma^2/2)(T-t)]/(\sigma\sqrt{T-t}) \tag{1.16}$$

and $d_2 = d_1 - (\sigma\sqrt{T-t})$.

This formula opened the gates to a flood of research on option pricing theory, since it is an exact formula for an asset price, based on observable variables. This is a rare finding from a general equilibrium model. It is still the goal in option pricing to derive a formula based on obervable variables similar to the Black-Scholes model under different conditions such as stochastic interest rates and volatility, and alternative price processes.

The most noticeable property of (1.15) is the variables that it does not depend on. This option price does not depend on the expected return μ of the underlying stock, risk preferences of investors, or on the aggregate supply of assets. It does depend on the rate of interest and the volatility of the return σ on the underlying stock. At the beginning of the study of options, volatility was considered to be a stable number, so they could estimate it from time series data. However, this is usually not the case in practise. It has been known that historical data do not provide a good estimate of volatility parameter. Currently, one way of estimating the volatility parameter is by implied volatility (see Hull [21] for further details).

It should be noted that the Samuelson and Merton model [52] is also a complete general equilibrium formulation, and is very simple (three assets and one investor). Their formula is

$$c(S_t, T-t; X) = e^{-r(T-t)} \int_{X/S_t}^{\infty} (zS_t - X) dQ(z; T-t), \qquad (1.17)$$

where dQ is a probability density function with $\int_{-\infty}^{\infty} z\, dQ(z; T-t) = e^{r(T-t)}$. Equations (1.15) and (1.17) will be the same only when dQ is a lognormal density and

$$\int_{-\infty}^{\infty} (\log z)^2\, dQ(z; T-t) - \left(\int_{-\infty}^{\infty} \log z\, dQ(z; T-t) \right)^2 = \sigma^2 (T-t).$$

Note that this occurs only if

1. the objective returns on the stock are lognormally distributed.

2. the investor's utility function is iso-elastic. That is, the utility function is assumed to be of a form yielding constant relative risk aversion (see [42]).

However, dQ is a risk-adjusted distribution, dependent on both risk preferences and aggregate supplies, while the distribution in (1.15) is the objective distribution of returns on the underlying stock. Black and Scholes claim one possible reason that Samuelson and Merton [52] did not arrive at formula (1.15) was because they did not consider other assets. However, it is necessary to consider only three assets to derive the Black-Scholes formula (see [44]).

The Black and Scholes derivation of (1.15) is intuitively appealing, but a rigorous derivation can also be provided, particularly to gain an insight into the sufficient conditions for the formula to obtain. Note that because Black and Scholes consider only terminal boundary conditions, their analysis is only applicable to European options, even though the European valuation is often equal to the American option. Finally, although their model is based on a different economic structure, the formal analytical content is identical to Samuelson's [50] linear $\alpha = \beta$ model when the returns on the underlying stock are lognormal. Hence, with a different interpretation of the parameters, theorems proved in Samuelson [50] and McKean [41] are directly applicable to the Black-Scholes models and vice versa.

1.3 Alternative derivation

For the alternative derivation of Black-Scholes formula, we consider the case of a European option where no payouts are made to the underlying stock over the life of the contract. We make the following further assumptions. For further details, see Merton [44].

Assumption 3 *"Frictionless" market: There are no transaction costs or differential taxes. Trading takes place continuously and borrowing and short-selling are allowed without restriction, and the borrowing rate equals the lending rate.*

Assumption 4 *Stock price dynamics: The instantaneous return on the underlying stock $S = \{S_t : 0 \leq t \leq T\}$ is described by the stochastic differential equation*

$$\frac{dS}{S} = \mu dt + \sigma dW \tag{1.18}$$

where, as before (see (1.14)), μ is the expected return on the underlying stock, σ^2 is the instantanous variance of the return and $W = \{W_t : 0 \leq t \leq T\}$ is standard Brownian motion. μ is a random variable which may even be dependent on the level of the stock price or other assets' returns. $\sigma = \sigma_t$ is a non-random, known function of time.

The solution of (1.18) is

$$S_t = S_0 e^{\sigma W_t + (\mu - \sigma^2/2)t}$$

where S_0 is the stock price process at time 0.

REMARK: Note that S is neither an independent increment process nor stationary, although W clearly is.

Assumption 5 *Bond price dynamics: With $B(T)$ as before,*

$$\frac{dB}{B} = \alpha(T)dt + \delta(T)d\widetilde{W}(t,T) \tag{1.19}$$

where α is the instantaneous expected return, δ^2 is the instantaneous variance, and $\widetilde{W} = \{\widetilde{W}(t,T) : 0 \leq t \leq T\}$ is the standard Brownian motion for a riskless bond with maturity date T.

This is consistent with the general efficient market hypothesis of Fama [14] and Samuelson [51]. The coefficient $\alpha(T)$ may be random, and different for different maturities. Because $B(T-t)$ is the price of a discounted loan with no risk of default, $B(0) = 1$ with certainty, and $\delta(0) = 0$. However, δ is non-stochastic and independent of B. In the special case when the interest rate is constant over time, $\delta = 0$, $\alpha = r$, and $B(T-t) = e^{-r(T-t)}$.

Assumption 6 *Investor's preference and expectations: Apart from Assumption 2, it is assumed that all investors agree on the values of σ and δ, and on the distributions of W and $\widetilde{W}$. It is not assumed that they agree on either μ or α.*

Recall that we assume that the option price is a function of the stock price, the riskless bond, and the length of time to expiration. If we assume the distributions of S and B and we consider any option price as a function $H(S_t, B, T-t; X)$, then by Itô's lemma (also suppressing the subscript), we have

$$\begin{aligned} dH &= \frac{\partial H}{\partial S} dS + \frac{\partial H}{\partial B} dB + \frac{\partial H}{\partial T} dT \\ &+ \frac{1}{2}\left[\frac{\partial^2 H}{\partial S^2}(dS)^2 + 2\frac{\partial^2 H}{\partial S \partial B}(dSdB) + \frac{\partial^2 H}{\partial B^2}(dB)^2\right], \qquad (1.20) \end{aligned}$$

where $(dS)^2 \equiv \sigma^2 S^2 dt$, $(dB)^2 \equiv \delta^2 B^2 dt$, $dT = -dt$, and $(dSdB) \equiv \rho\sigma\delta SBdt$ with ρ the instantaneous correlation coefficient between the (unanticipated) return on the stock and on the bond. So, the change in the option price over time satisfies the stochastic differential equation (1.20). Substituting (1.18) and (1.19) and re-arranging terms, (1.20) becomes

$$dH = \beta H dt + \gamma H dW + \eta H d\widetilde{W}, \qquad (1.21)$$

where $\beta = [\sigma^2 S^2(\partial^2 H/\partial S^2)/2 + \rho\sigma\delta S\, B(\partial^2 H/\partial S\, \partial B) + \delta^2 B^2 (\partial^2 H/\partial B^2)/2 + \alpha B(\partial H/\partial B) + \mu S(\partial H/\partial S) - (\partial H/\partial(T-t))]/H$, $\gamma =$

$\sigma S(\partial H/\partial S)/H$, and $\eta = \delta B(\partial H/\partial B)/H$. Denoting by u_1(resp. u_2, u_3) the amounts invested in the underlying stock (resp., option, bonds), it can be shown (Merton [42] or [43]) that the instantaneous return on investment is

$$[(\mu-\alpha)u_1+(\beta-\alpha)u_2]dt+[\sigma u_1+\gamma u_2]dW+[\eta u_2-\delta(u_1+u_2)]d\widetilde{W}. \quad (1.22)$$

To make the above return non-stochastic, the coefficients of dW and $d\widetilde{W}$ in (1.22) must be zero. To avoid arbitrage, the coefficient of dt must then be zero. This gives a system of three equations in u_1 and u_2, which will have a non-zero solution if and only if

$$\begin{aligned} \sigma(\beta-\alpha) &= \gamma(\mu-\alpha) \\ \delta\gamma &= \sigma(\delta-\eta). \end{aligned} \quad (1.23)$$

When r is constant, we have $\delta = 0$, $\alpha = r$, and $B(T-t) = e^{-r(T-t)}$. Thus (1.23) reduces to

$$\beta - r = \frac{\gamma}{\sigma}(\mu - r). \quad (1.24)$$

Making use of (1.24) and the fact that $B(T-t)$ is non-random for this model of the American put P_t, we obtain

$$\frac{1}{2}\sigma^2 S^2 \frac{\partial^2 P}{\partial S^2} + rS\frac{\partial P}{\partial S} - rP - \frac{\partial P}{\partial T} = 0, \quad (1.25)$$

subject to $P(\infty, T, X) = 0$, $P(S_T, 0, X) = (X - S_T)^+$, and $P(S_t, T - t, X) \geq (X - S_t)^+$.

For $T = \infty$, equation (1.25) is an ordinary differential equation

$$\frac{1}{2}\sigma^2 S^2 \frac{d^2 P}{dS^2} + rS\frac{dP}{dS} - rP = 0, \quad (1.26)$$

valid for all $S \geq S_c$, with the boundary conditions

$$P(\infty, \infty, X) = 0,$$

$$P(S_c, \infty, X) = X - S_c,$$

and S_c chosen so as to maximize the return. It is elementary that the solution of (1.26), subject to the above boundary conditions, is

$$P(S, \infty, X) = (X - S_c)(S/S_c)^{-M}. \tag{1.27}$$

The value of S_c which maximizes (1.27) is easily found to be $S_c = MX/(1+M)$, whence (1.27) becomes

$$P(S_t, \infty, X) = \frac{X}{1+M}\left(\frac{(1+M)S_t}{MX}\right)^{-M}. \tag{1.28}$$

Chapter 2
Optimal Stopping and the American Put

An American put option provides *the right, but not the obligation, to sell for price X a unit of stock at any time up to the time horizon T*, the original expiry date of the option. To simplify the problem, we follow the Black-Scholes model by assuming that the stock pays no dividends during the lifetime of the option and that the stock price S_t follows geometric Brownian motion:

$$S_t = S_0 e^{(\sigma W_t + (\mu - \sigma^2/2)t)} \tag{2.1}$$

where $(W_t)_{t\geq 0}$ is a standard Brownian motion, S_0 $(= x)$ is the fixed initial value, μ is the expected rate of return on the stock, and σ is the variance of the return of the stock. This price S_t is the unique (strong) solution of the stochastic differential equation

$$dS_t = S_t(\sigma dW_t + \mu dt). \tag{2.2}$$

We also assume that interest on cash is earned at a fixed rate $r > 0$.

In 1991, Jacka established the following basic results: (i) The fair price for an option is a function of the present stock price and the time horizon only, and is the unique solution of a parabolic free boundary problem. (ii) The option price is the solution of an optimal stopping problem which is a generalization of the Black-Scholes option pricing formula. These results are well-known. In particular, Karatzas [28] proved (ii). McKean [41] and Van Moerbeke [55] had earlier derived a result equivalent to (i), and parallel results were established by Kim [33], Carr, Jarrow and Myneni [7], and Jacka and Lynn [24]. In this chapter, we present Jacka's paper on the optimal stopping and the American put (see Jacka [22]).

2.1 The Optimal Stopping Problem

We consider the problem of optimally stopping $e^{-rt}S_t$ before time T, that is, we find

$$f(x,t) \stackrel{\text{def}}{=} \sup_{\tau \le t} E_x[e^{-r\tau}(X - S_\tau)^+] \tag{2.3}$$

with τ ranging over stopping times. Let P be a probability measure such that the discounted stock price $e^{-rt}S_t$ is a martingale with respect to P. It will be shown (Theorem 8) that there is an exercise boundary $b(t) = S_c$ and that we exercise the option the first time that S_s, the price of the underlying stock, falls below the (lower) boundary $b(t)$ at $t = T - s$.

It is well known that under P, S_t satisfies the stochastic integral equation

$$S_t(x) = x + \int_0^t \sigma S_s(x) dW_s + \int_0^t r S_s(x) ds \tag{2.4}$$

where W is the P-Brownian motion so that

$$S_t(x) = x e^{(\sigma W_t + (r - \sigma^2/2)t)}. \tag{2.5}$$

Let $g(x) = (X - x)^+$. Then one has the following:

Theorem 3 *The payoff function V_T for the stopping problem is a function of the present price of the stock and of the time to the expiry of the option only. That is,*

$$V_T = f(S_0, T). \tag{2.6}$$

The function f is continuous, $f(x,t) \geq g(x)$, and the optimal stopping time τ^* is

$$\tau^* = \inf\{s : f(S_s, T-s) = g(S_s)\} = \inf\{t \geq 0 : (S_t, T-t) \notin D\} \tag{2.7}$$

where

$$D = \{(x,t) \in \mathbf{R}^+ \times \mathbf{R}^+ : f(x,t) > g(x)\}$$

is the continuation region for the stopping problem.

PROOF. The results follow from Krylov's theorems (Appendix, Theorems 9 and 10) upon letting $w(t,x) = f(x,t)$ in Krylov's notation. □

Further, one knows the following about the structure of the region D:

Proposition 1 *For each $t > 0$, the t-section of D has the form*

$$D_t \equiv \{x : (x,t) \in D\} = (b(t), \infty)$$

for some $b(t)$ satisfying $X > b(t) > 0$.

PROOF. We know that zero is not in D_t, so we only need to show that for any $y > x$, $(x \in D_t) \Longrightarrow (y \in D_t)$. We show this by using a pathwise comparison of $S_t(x) = xe^{(\sigma W_t + (r-\sigma^2/2)t)}$. Suppose $x \in D_t$, $y > x$, and let $\tau = \inf(s \geq 0 : (S_s(x), T-s) \notin D)$. (That is, τ is a stopping time for $S_s(x)$.) Note that τ is bounded. Then

$$
\begin{aligned}
&f(y,t) - f(x,t) \\
&\quad = f(y,t) - E[e^{-r\tau}(X - S_\tau(x))^+] \\
&\quad \geq E[e^{-r\tau}((X - S_\tau(y))^+ - (X - S_\tau(x))^+)] \\
&\quad = Ee^{-r\tau}[(X - S_\tau(y)) - (X - S_\tau(x)) \\
&\qquad + (X - S_\tau(y))^- - (X - S_\tau(x))^-]. \qquad (2.8)
\end{aligned}
$$

Now $S_\tau(y) > S_\tau(x)$ since $y > x$, so the last term of (2.8) is non-negative. Thus

$$
\begin{aligned}
f(y,t) - f(x,t) &\geq E[e^{-r\tau}(S_\tau(x) - S_\tau(y))] \\
&= (x - y)E[e^{(\sigma W_\tau - \sigma^2\tau/2)}] = (x - y),
\end{aligned}
$$

since $e^{(\sigma W_t - \sigma^2 t/2)}$ is a martingale starting at 1 and τ is feasible for $S_s(y)$. Therefore,

$$f(y,t) \geq f(x,t) + (x - y).$$

If $f(x,t) > (X - x)^+$, then

$$f(y,t) > (X - x)^+ + (x - y) \geq (X - x) + (x - y) = (X - y).$$

Since $f(y,t) > 0$, for all $t > 0$ and all $y \geq 0$ (Appendix, Lemma 2.16), we have $f(y,t) > g(y)$, so $(y,t) \in D$, as required. So D_t is an interval (possibly closed). Now let $x \in D_t$. It follows because $f(x,t) > 0$ for every x, t (since f is continuous) that D_t is the open interval $(b(t), \infty)$ with $0 < b(t) < X$. □

Note that $b(t) \leq X$ for all $t \geq 0$, since if $x > X$, $(X - x)^+ = 0$ but $f(x,t) > 0$. It can be shown (see Appendix, Propositions 3 and 4) that b is a continuous, decreasing function of t.

The limiting behaviour of b and f is given by:

Proposition 2 *Let (F, b) be the unique solution, with $F \in C^1$, of*

$$\begin{aligned} \sigma^2 x^2 F''(x)/2 + rxF'(x) - rF = & \quad 0 \quad \text{if } x \geq b \\ F(x) = & \quad X - x \quad \text{if } x < b \end{aligned} \tag{2.9}$$

Then

(i) $b = [MX/(1+M)]$,

$$F(x) = \begin{cases} (X-b)(b/x)^M & \text{if } x \geq b \\ (X-x) & \text{if } x < b \end{cases}$$

where $M = 2r/\sigma^2$ *and* $F(x) = \sup_\tau \; E_x[e^{-r\tau} g(S_\tau)]$.

(ii) $\lim_{t\to\infty} f(x,t) = F(x)$.

(iii) $\lim_{t\to\infty} b(t) = b$.

PROOF. To begin with, observe that f is decreasing in x, as it is immediate from its definition. Also, f is increasing in t, since if τ is admissible for the optimal stopping problem with horizon t, it is also admissible for the problem with horizon s, for any $s \geq t$.

(i) One can see, for example, Merton [44] or Karatzas and Shreve, [32] for a proof of this. We will provide a sketch of the proof.

From the theory of linear ordinary differential equations, the solutions of (2.9) involves two constants a_1 and a_2. The boundary conditions will determine these constants along this unknown lower bound b. Recall that the boundary conditions for F are

$$F(\infty) = 0 \tag{2.10}$$

and

$$F(b) = X - b. \tag{2.11}$$

The general solution of (2.9) is

$$F(x) = a_1 x + a_2 x^{-M},$$

where $M = 2r/\sigma^2$. Equation (2.10) requires that $a_1 = 0$ and (2.11) requires that $a_2 = (X - b)b^M$. Hence, we have

$$F(x) = (X - b)\left(\frac{x}{b}\right)^{-M} \tag{2.12}$$

To determine b we must find the value of b which maximizes (2.12). That is, choose $b = b^*$ such that $\frac{\partial F}{\partial b^*} = 0$ and $\frac{\partial^2 F}{\partial b^{*2}} < 0$. From this, we have $b = [MX/(1 + M)]$ as desired.

(ii) It is clear that $f(x,t) \le F(x)$ for all t. Conversely, let $\tau = \inf\{t : S_t \le b\}$. Then $F(x) = E_x[e^{-r\tau} g(S_\tau(x))]$. If we let $\tau_t = \tau \wedge t$, then for $x > b$,

$$\begin{aligned} f(x,t) &\ge E_x\left[e^{-r\tau_t} g(S_{\tau_t})\right] \\ &\ge (X - b)E_x\left[e^{-r\tau_t}\mathbf{1}_{\tau \le t}\right] \\ &\longrightarrow (X - b)E_x[e^{-r\tau}] \equiv F(x) \end{aligned}$$

as $t \to \infty$, since $e^{-r\tau}$ is a bounded continuous function on $\mathbf{R}^+$, and $\tau_t \to \tau$. On the other hand, for $x \le b$, we have $f(x,t) \ge g(x) = F(x)$, so $\lim_{t\to\infty} f(x,t) = F(x)$.

(iii) Given $x > b$, we have $F(x) > g(x)$. Let $\varepsilon = F(x) - g(x)$, and take $t(\varepsilon)$ such that

$$F(x) - f(x,t) \le \varepsilon/2$$

for all $t \ge t(\varepsilon)$. Then $f(x,t) > g(x)$ for $t \ge t(\varepsilon)$, so $(x,t) \in D$, whence $b(t) \le x$. But this is true for all $x > b$, so $\limsup_{t\to\infty} b(t) \le b$.

Similarly, if $x < b$, then $g(x) = F(x) \geq f(x,t)$, so $x \leq b(t)$ for all t, and so $b \leq \liminf_{t\to\infty} b(t)$. The result follows. □

REMARK. If one differentiates the solution in Proposition 2 (i) with respect to x and evaluate at $x = b^*$, then we have the Samuelson "high contact" boundary condition (see Merton [46], p. 300). That is,

$$F'(b^*) = -1.$$

Jacka's main theorem is the following.

Theorem 4 *The pair (f,b) is a solution pair (φ,h) of the free boundary problem*

$L\varphi = 0,\ x > h(t)$

$\varphi(h(t),t) = X - h(t)$

$\frac{\partial\varphi}{\partial x}(h(t),t) = -1$

$\varphi(x,0) \to g(x),\ x \geq h(0)$

$\varphi(x,t) \to 0$ *uniformly on compact sets as* $x \to \infty$

and f is maximal.

Before proceeding with the proof of the above theorem, we need several lemmas. Recall that a function on a metric space X is Hölder continuous if, for all $x \in X$, there exist constants $C_x > 0$ and $0 < \gamma_x \leq 1$ and an open neighbourhood $\mathcal{V}$ of x such that $d(f(x_1), f(x_2)) \leq C_x d(x_1,x_2)^{\gamma_x}$ for all $x_1, x_2 \in \mathcal{V}$. C^1 and C^2 denote the class of functions with continuous first derivative and second derivative, respectively.

Lemma 1 *f is C^1 in x.*

PROOF. Define

$$M_s = M_s^{t,f} = e^{-rs} f(S_s, t-s) + rX \int_0^s e^{-ru} \mathbf{1}_{S_u \leq h(t-u)} du. \qquad (2.13)$$

Since $\partial D = \{b(s,s); s \geq 0\}$ has zero measure, and the decreasing component of the semimartingale decomposition of $(e^{-rt}g(S_t))$ is absolutely continuous with respect to the Lebesgue measure, it follows from Theorem 12 and Corollary 3 of the Appendix, that $M_s^{t,f}$ is a martingale. Therefore, we can write

$$f(x,t) = M_0 = EM_t = p(x,t) + rX \int_0^t e^{-ru}\psi(x, b(t-u), u)du,$$

where $p(x,t) = E(e^{-rt}(X - S_t)^+)$ (price of an European put option) and $\psi(x,y,t) = \Pr(S_t(x) \leq y)$. Then we can rewrite $\frac{\partial f}{\partial x}(x,t) \equiv f_x(x,t)$ as

$$f_x(x,t) = rX \int_0^t e^{-rs}\psi_x(x, b(t-s), s)ds + p_x(x,t)$$

(Lebesgue a.e.).

Define $u(x,t) = \int_0^t e^{-rs}\psi_x(x, b(t-s), s)ds$. We need to show that $u(x,t)$ is continuous. Since b is bounded and Borel measurable, it is sufficient to show that for any Borel measurable function k, the function

$$d(x,t) = \int_0^t e^{-rs} n(x - k(s), \sigma^2 s)ds$$

where $n(x,v) = \frac{1}{\sqrt{2\pi v}} e^{\frac{-x^2}{2v}}$ is the normal density at x with mean zero and variance v, is continuous in x. Clearly, for any $0 \leq \delta \leq t$, we have $| n_x | \leq (e/2\pi)^{1/2}/(\sigma^2 t)$ and $0 < n < 1/(2\pi\sigma^2 t)^{1/2}$, so

$$\begin{aligned}
&| d(x,t) - d(y,t) | \\
&= \left| \int_0^t e^{-rs} n(x - k(s), \sigma^2 s)ds - \int_0^t e^{-rs} n(y - k(s), \sigma^2 s)ds \right| \\
&\leq \int_0^t | e^{-rs} n(x - k(s), \sigma^2 s) | ds + \int_0^t | e^{-rs} n(y - k(s), \sigma^2 s) | ds \\
&\leq \int_0^\delta ds/\sigma(2\pi s)^{1/2} + \int_\delta^t | x - y | (e/2\pi)^{1/2}/(\sigma^2 s)ds \\
&= (2\sqrt{\delta})/(\sigma\sqrt{2\pi}) + | x - y | (e/2\pi)^{1/2} \log(t/\delta)/\sigma^2.
\end{aligned}$$

In particular, letting $\delta = (x-y)^2$, we get

$$\begin{aligned}
&| \, d(x,t) - d(y,t) \, | \\
&\quad \leq 2 \mid x-y \mid /(\sigma\sqrt{2\pi}) + \mid x-y \mid (e/2\pi)^{1/2}/\sigma^2(\log(t/\mid x-y \mid^2)) \\
&\quad = \mid x-y \mid [(2/\sigma^2)(e/2\pi)^{1/2}\{\sigma e^{-1/2} + (1/2)\log t - \log \mid x-y \mid\}] \\
&\quad \leq K_t \mid x-y \mid (1+ \mid \log \mid x-y \mid \mid).
\end{aligned}$$

Therefore, $| \, d(x,t) - d(y,t) \, | \leq K_t \mid x-y \mid (1+ \mid \log \mid x-y \mid \mid)$, so d is locally Hölder continuous. Thus, f is in C^1. □

Lemma 2 *Let* $\mathcal{L} = \frac{\sigma^2 x^2}{2}\frac{\partial^2}{\partial x^2} + rx\frac{\partial}{\partial x} - r - \frac{\partial}{\partial t}$ *(a parabolic operator). Then f is the unique solution φ of the initial value problem*

$\mathcal{L}\varphi = 0, \; x > b(t)$
$\varphi(b(t),t) = X - b(t)$, *where* $t \geq 0$
$\varphi(x,0) = g(x)$
$\varphi(x,t) \to 0$ *as* $x \to \infty$, *uniformly on compact sets.*

PROOF. Take $(x,t) \in D$. D is open so we can take an open rectangle $R = (x_1,x_2) \times (t_1,t_2)$ with $(x,t) \in R \subseteq D$. Consider the initial value problem $\mathcal{L}\varphi = 0$ in R and $\varphi(x,t) = f(x,t)$ on $\partial R \setminus (x_1,x_2) \times \{t_2\}$. The existence and uniqueness of a solution φ to this problem follows from Theorem 11 of the Appendix. Next, let

$$N_s^t = e^{-rs}\varphi(S_s(x), t-s).$$

It follows from Itô's formula that $N_{s\wedge\tau}$, where $\tau = \inf\{s \geq 0 : (S_s, t-s) \notin R\}$, is a bounded martingale, so that,

$$\varphi(x,t) = N_0^t = EN_\tau^t = Ee^{-r\tau}f(S_\tau, \tau - t) = f(x,t),$$

since τ is bounded from above by the first exit time of $(S_s(x), t-s)$ from D. Therefore f satisfies $\mathcal{L}f = 0$ in D.

The boundary conditions are also satisfied by f. The fact that $f(x,t) \to 0$ as $x \to \infty$ uniformly on compact sets follows, since $f(x,t) \to 0$ as $x \to \infty$ for each t ($0 < f(x,t) \leq X \Pr(\tau_X(S.(x)) \leq t)$ for $x > X$, and $\tau_X(S.(x)) \to \infty$ as $x \to \infty$ a.s.) and f is increasing in t and decreasing in x, as noted at the beginning of the proof of Proposition 2. Recall that $\tau_X(S.) \equiv \inf\{s \geq 0 : S_s \leq X\} \wedge t$. As for uniqueness, for any T, φ is bounded on $\mathbf{R}^+ \times [0,T]$ and $\varphi(x,t) = (X - x)$ for $x \leq b(t)$. Define

$$M_t^T = e^{-r(t\wedge\tau)}\varphi(S_{t\wedge\tau}, T - (t \wedge \tau))$$

where $\tau = \inf\{s \geq 0 : S_s = b(s)\}$ and $S_0 = x \geq b(T)$. By Itô's formula, M_t^T is a bounded martingale, so we have

$$M_0^T \equiv \varphi(x,T) = EM_T^T = Ee^{-r(T\wedge\tau)}g(S_{T\wedge\tau}) = f(x,T),$$

for all $T \in \mathbf{R}^+$, since we know $T \wedge \tau$ is optimal for the T-horizon problem, as required. □

Proof of Theorem 4. That (f, b) satisfies the conditions follows immediately from Lemmata 1 and 2. It remains to show that f is maximal. Suppose (φ, h) is another solution pair. By the same argument as in Lemma 2, we get $\varphi(x,t) \leq E(e^{-r(t\wedge\tau_h)}g(S_{t\wedge\tau_h}))$ since $g(x) \geq (X - x)$ and $\tau_h = \inf\{s \geq 0 : S_s = h(t - s)\}$. Thus $\varphi(x,t) \leq f(x,t) = \sup_{\tau\leq t} Ee^{-r\tau}g(S_\tau)$, as required. □

2.2 Pricing the Option

The problem is to find a fair price for an American put option. The next theorem is proved by Karatzas [28]. For further details, the reader should refer to this paper. Recall that S_t is defined by (2.1).

Theorem 5 *Let $dS'_t = \sigma S'_t dW_t + rS'_t dt$ with $S'_0 = S_0$ and let $f(x,t) = \sup_{\tau \le t} E_x[e^{-r\tau}(X - S'_\tau)^+]$ be the optimal payoff from optimally stopping the process $e^{-rs}(X - S_s)^+$. Then there is a self-financing portfolio of stock and cash, $((Y_s, Z_s); 0 \le s \le t)$ with Y and Z adapted, and whose value V_s at time s satisfies $V_s \equiv (S_s Y_s + Z_s) \ge (X - S_s)^+$ and such that if τ^* is the optimal stopping time corresponding to the payoff function $f(S_0, t)$, then $V_{\tau^*} = (X - S_{\tau^*})^+$. Therefore $S_0 Y_0 + Z_0 \equiv f(S_0, t)$ is the fair price for the option.*

That is, if one writes the option and sells it for V_0 and then purchase the portfolio (S_0, Y_0) and manages it in the prescribed manner, then one will not lose money, and will make money unless the option is not rationally used.

PROOF OF THEOREM 5. In the previous section it was shown that the optimal stopping time t is to exercise the option when S'_s falls below $b(t-s)$. Moreover f is piecewise C^2 in x, C^1 in t with

$$d(e^{-rs} f(S'_s, t-s)) = e^{-rs} f_x(S'_s, t-s)\sigma S'_s dW_s - rXe^{-rs}\mathbf{1}_{S'_s \le b(t-s)} ds.$$

So

$$\begin{aligned} df(S'_s, t-s) &= rf(S'_s, t-s)ds + f_x(S'_s, t-s)\sigma S'_s dW_s \\ &\quad - rX\mathbf{1}_{S'_s \le b(t-s)} ds. \end{aligned}$$

Set

$$\begin{aligned} Y_s &= f_x(S_s, t-s), \\ Z_s &= f(S_s, t-s) - S_s f_x(S_s, t-s) + rXe^{rs}\int_0^s e^{-ru}\mathbf{1}_{S_u \le b(t-u)} du, \\ V_s &\equiv S_s Y_s + Z_s \\ &= S_s f_x(S_s, t-s) + f(S_s, t-s) - S_s f_x(S_s, t-s) \end{aligned}$$

$$
\begin{aligned}
& + rXe^{rs}\int_0^s e^{-ru}\mathbf{1}_{S_u\le b(t-u)}du \\
= \quad & f(S_s, t-s) + rXe^{rs}\int_0^s e^{-ru}\mathbf{1}_{S_u\le b(t-u)}du.
\end{aligned}
$$

Then

$$
\begin{aligned}
dV_s \quad = \quad & rf(S_s,t-s)ds + \sigma f_x(S_s,t-s)S_s dW_s \\
& +(\mu - r)f_x(S_s,t-s)S_s ds + r\left(e^{rs}rX\int_0^s e^{-ru}\mathbf{1}_{S_u\le b(t-u)}du\right)ds \\
= \quad & Y_s dS_s + rZ_s ds.
\end{aligned}
$$

Since Y_s, S_s, and Z_s are continuous, V is the value of a self-financing portfolio, so $dV_s = Y_{S}\text{-}dS_s + rZ_{S}\text{-}dS_s$ is the instantaneous return of a portfolio of $Y_{S}-$ shares and $Z_{S}-$ cash. By $V_s \equiv S_sY_s + Z_s$, we know that $V_t \ge (X - S_t)^+$ since $f(x,t) \ge g(x)$. □

2.3 Integral Transform of the Stopping Time

In this section, we see that we can define a martingale for every solution of the free boundary problem (see Theorem 6). In Theorem 7, we have another characterization of (f, b).

Theorem 6 *Let (φ,h) be a solution pair of the free boundary problem in Theorem 4. Then*

$$
M_s \equiv M_s^{t,\varphi} = e^{-rs}\varphi(S_s, t-s) + rX\int_0^t e^{-ru}\mathbf{1}_{S_u\le h(t-u)}du, \qquad (2.14)
$$

is a martingale.

PROOF. By Itô's formula,

$$\begin{aligned} dM_s &= e^{-rs}(\mathcal{L}\varphi)(S_s, t-s)ds + rXe^{-rs}\mathbf{1}_{S_s \le h(t-s)}ds \\ &\quad + e^{-rs}\varphi(S_s, t-s)\sigma S_s dW_s \end{aligned}$$

since φ is C^1, piecewise C^2 in x and piecewise C^1 in t. But

$$\mathcal{L}\varphi(x,s) = \begin{cases} 0 & \text{if } x > h(s) \\ -rX & \text{if } x < h(s) \end{cases}$$

so M is a local martingale. Since $\varphi(x,u) \to 0$ as $x \to \infty$ uniformly for u in a compact set, and φ is C^1 in x, φ is bounded on $\mathbf{R}^+ \times [0,t]$, so M is a martingale. □

Theorem 7 *The pair (f,b) is the unique solution pair (φ,h) of the free boundary problem of Theorem 4 (with h Lebesgue measurable), satisfying $h(t) > 0$ infinitely often.*

PROOF. We know that (f,b) satisfies the conditions of Theorem 4. Now suppose (φ,h) is another solution pair of (f,b) with $h(t) > 0$ infinitely often. Define

$$M_s^0 = e^{-rs}f(S_s(x), t-s) + rX \int_0^s e^{-ru}\mathbf{1}_{S_u \le b(t-u)}du$$

and

$$M_s^1 = e^{-rs}\varphi(S_s(x), t-s) + rX \int_0^s e^{-ru}\mathbf{1}_{S_u \le h(t-u)}du.$$

For a fixed $t \ge 0$, consider $x < b(\infty) \wedge h(t) \le b(t) \wedge h(t)$. The last inequality holds since b is bounded by X and b is decreasing in t (see Appendix, Proposition 3). Thus, $x < b(t) \wedge h(t)$, $f(x,t) = \varphi(x,t) = g(x)$. Define $M = M^0 - M^1$ with $M_0 = 0$. Define $\tau = \inf\{s \ge 0 : S_u > b(t-u)\} \wedge t$. Then $EM_{\tau\wedge s} = M_0 = 0$, but $M_{\tau\wedge s} = e^{-r(\tau\wedge s)}[f(S_{\tau\wedge s}(x), t-(\tau\wedge s)) - \varphi(S_{\tau\wedge s}(x), t-(\tau\wedge s))] + rX \int_0^{\tau\wedge s} e^{-ru}\mathbf{1}_{S_u(x) > h(t-u)}du$. Then the right hand side is positive because f is maximal and $S_u(x) \le b(t-u)$ on $[0,\tau]$. Therefore, $S_{\cdot\wedge\tau\wedge s}(x) \le h(t-(\cdot\wedge\tau))$ (*Lebesgue* $\times$ *P*) a.e.

and $\varphi(S_{s\wedge\tau}(x), t-s) = f(S_{s\wedge\tau}(x), t-s) = g(S_{s\wedge\tau})$ (P a.s.). Note that $h(s) \leq b(s)$ for $s \leq t$.

Next let

$$M_t = rX \int_0^t e^{-rs}\{\mathbf{1}_{S_s \leq b(t-s)} - \mathbf{1}_{S_s \leq h(t-s)}\}ds,$$

and let $G = \mathbf{1}_{S_s \leq b(t-s)} - \mathbf{1}_{S_s \leq h(t-s)}$. Then G is negative since $\varphi(x,0) = f(x,0) = g(x)$. But $EM_t = M_0 = 0$, so G is zero (*Lebesgue* $\times$ P) almost everywhere (a.e.). Therefore, $h = b$ a.e. on $[0,t]$. For arbitrary x and $s \leq t$,

$$\varphi(x,s) = M_0^{s,1} = EM_s^{s,1} = M_0^{s,0} = f(x,s),$$

so $\varphi = f$ and hence $h \leq b$ everywhere else on $\{s \in [0,t]\}$ since $f > g$ on D. This implies that $h = b$ since $\mathcal{L}f \neq 0$ on D^c. □

Note that the argument fails if there exist a t such that $h(s) \equiv 0$ $\forall s \geq t$. Also, the theorem indicates that prohibited stopping before the horizon t corresponds to the case where $h(s) = 0$ for all $s > t$.

One also has the following result which was also noted by El Karoui and Karatzas [13].

Theorem 8 *The boundary $b(\cdot)$ is the unique left continuous solution $h(\cdot)$, satisfying $X > h > 0$ for all $t > 0$, of the integral equation*

$$X - x = p(x,t) + rX \int_0^t \psi(x, h(t-s); s)ds \qquad (2.15)$$

for all $x \leq h(t)$, where $p(x,t) = E[e^{-rt}(X - S_t(x))^+]$ is the price of the European option, and $\psi(x,y;t) = \Pr(S_t(x) \leq y)$.

PROOF. Since

$$M_s^0 \equiv e^{-rs} f(S_s(x), t-s) + rX \int_0^s e^{-ru} \mathbf{1}_{S_u(x) \leq b(t-u)} du,$$

for $s \leq t$, is a martingale, with initial value $(X - x)$ if $x \leq b(t)$, the boundary b satisfies the conditions. Suppose h also satisfies the conditions. Define

$$\varphi(x,t) = p(x,t) + rX \int_0^t e^{-rs}\psi(x, h(t-s), s)ds.$$

Then we can write

$$M_s^1 \equiv e^{-rs}\varphi(S_s(x), t-s) + rX \int_0^s e^{-ru}\mathbf{1}_{S_u(x)\leq h(t-u)}du,$$

which is a $(\sigma W_u : u \leq s)$ martingale for $(x,t) \in \mathbf{R}_+^2$. Define

$$\tau = \inf\{s \geq 0 : S_s(x) \leq h(t-s)\} \wedge t.$$

For $x \geq h(t)$, let

$$\varphi(x,t) \equiv M_0^1 = Ee^{-r\tau}\varphi(S_\tau, t-\tau) = Ee^{-r\tau}g(S_\tau).$$

Then $\varphi \leq f$ if $x \geq h(t)$ and $\varphi = g \leq f$ if $x \leq h(t)$. Now take $0 < x \leq h(t) \wedge b(t)$ and set

$$\theta = \inf\{s \geq 0 : S_s(x) \geq b(t-s)\} \wedge t.$$

Then

$$\begin{aligned} 0 &= f(x,t) - \varphi(x,t) \\ &= Ee^{-r\theta}[f(S_\theta, t-\theta) - \varphi(S_\theta, t-\theta)] \\ &+ ErX \int_0^\theta e^{-ru}1_{S_u > h(t-s)}du. \end{aligned}$$

Both terms on the right-hand side are nonnegative (since f is maximal), so $h \geq b$ (Lebesgue a.e). However this implies that $\varphi \geq f$. Therefore $\varphi = f$ and so (by the left continuity of b) one obtains $h = b$. □

The above theorem allows to find $b(\cdot)$ by solving the integral equation (2.15). We can do this by choosing the one that is left continuous from all possible solutions.

2.4 Appendix

In this section, we have collected some background information necessary for the previous sections. It should be noted that Krylov's results (below) are more general than needed to establish Jacka's results.

1. Krylov [35]

Let A be a separable metric space (a set of admissible controls); $\mathbf{R}^d$ the Euclidean space of dimension d; T a nonnegative number. Consider a controlled process in the space $\mathbf{R}^d$ in a time interval $[0, T]$. Let $(W_t, \mathcal{F}_t)$ be a d_1-dimensional Brownian motion for some integer d_1. Suppose $\alpha \in A$, $t \geq 0$, $x \in \mathbf{R}^{\mathrm{d}}$, and suppose we are given $\sigma(\alpha, t, x)$ which characterizes the diffusion component of the process as a matrix of dimension $d \times d_1$ and $b(\alpha, t, x)$ which characterizes the deterministic component as a d-dimensional vector. We are also given real-valued functions $C^\alpha(t, x) \geq 0$, $f^\alpha(t, x)$, and $G(x)$, where $f^\alpha(t, x)\Delta t$ is the payoff during the time interval $[t, \Delta t + t]$, if the controlled process is near a point x at time t, and if a control α is used, $G(x)$ is the gain at time T, and $C^\alpha(t, x)$ is the measure of discounting.

Assume that σ, b, C, f, G are continuous with respect to (α, x) and continuous with respect to x uniformly over α for each t. Further, assume that the above functions are Borel measurable with respect to (α, t, x).

For some constants $m, k \geq 0$, all $x, y \in \mathbf{R}^{\mathrm{d}}$, $t \geq 0$, $\alpha \in A$, suppose

$$\| \sigma(\alpha, t, x) - \sigma(\alpha, t, y) \| + | b(\alpha, t, x) + b(\alpha, t, y) | \leq k \mid x - y \mid \quad (2.16)$$

$$\| \sigma(\alpha, t, x) \| + | b(\alpha, t, x) | \leq k(1 + | x |) \quad (2.17)$$

$$| C^\alpha(t, x) | + | f^\alpha(t, x) | + | G(x) | \leq k(1 + | x |)^m. \quad (2.18)$$

DEFINITION. A *strategy* means a process $\alpha_t(w)$ which is progressively measurable with respect to a system $\mathcal{F}_t$ of σ-algebras having values in A. Denote by $\mathcal{U}$ the set of all strategies.

For $s \le T$, let

$$\varphi_t^{\alpha,s,x} = \int_0^t C^{\alpha_r}(s+r, x_r^{\alpha,x,r})dr, \tag{2.19}$$

$$v^\alpha(s,x) = E[\int_0^{T-s} f^{\alpha_t}(s+t, x_t^{\alpha,s,x})e^{-\varphi_t^{\alpha,s,x}}\,dt + G(x_{T-s}^{\alpha,s,x})e^{-\varphi_{T-s}}] \tag{2.20}$$

and

$$v(s,x) = \sup_{\alpha\in\mathcal{U}} v^\alpha(s,x) \tag{2.21}$$

Note that the E in (2.20) refers to expectation.

Let $C([0,\infty),\mathbf{R^d})$ be the space of all continuous functions x_t with values in $\mathbf{R^d}$ defined on $[0,\infty)$, and N_t the smallest σ-algebra of subsets of $C([0,\infty),\mathbf{R^d})$ which contains all sets of the form $(x_{[0,\infty)} : x_r \in \Gamma)$ for $r \in [0,T]$ and Borel $\Gamma \subset \mathbf{R^d}$.

DEFINITION. For $t \in [0,\infty)$ and $x_{[0,\infty)} \in C([0,\infty),\mathbf{R^d}))$, an A-valued function α_t, with $\alpha_t(x_{[0,\infty)}) = \alpha_t(x_{[0,t)})$ is said to be a *natural strategy admissible at the point* (s,x) if α_t is progressively measurable with respect to (N_t) and if, in addition, there exist at least one solution of the stochastic equation

$$x_t = x + \int_0^t \sigma(\alpha_r(x_{[0,r)}), s+r, x_r)dW_r + \int_0^t b(\alpha_r(x_{[0,r)}), s+r, x_r)dr, \tag{2.22}$$

which is progressively measurable with respect to $\mathcal{F}_t$. Denote by $\mathcal{U}_E(s,x)$ the set of all natural strategies admissible at the point (s,x).

For each strategy $\alpha \in \mathcal{U}_E(s,x)$, choose one (fixed) solution $x_t^{\alpha,s,x}$ of (2.22).

DEFINITION. The natural strategy $\alpha_t(x_{[0,t)})$ is said to be a *(nonstationary) Markov strategy* if $\alpha_t(x_{[0,t)}) = \alpha_t(x_t)$ for some Borel function $\alpha_t(x_t)$. Denote by $\mathcal{U}_M$ the set of all Markov strategies admissible at the point (s,x). Note that $\mathcal{U}_M(s,x) \subset \mathcal{U}_E(s,x)$, and $\mathcal{U}_M(s,x) \neq \emptyset$.

Let $H(t,x)$ be a continuous function of (t,x), $(x \in \mathbf{R}^d, t \geq 0)$ such that

$$\mid H(t,x) \mid \leq k(1+ \mid x \mid)^m \tag{2.23}$$

for all t and x.

For $s \in [0,T]$, denote by $\mathcal{M}(T-s)$ the set of all Markov times (with respect to $\mathcal{F}_t$) not exceeding $(T-s)$. For $\alpha \in \mathcal{U}$, $\tau \in \mathcal{M}(T-s)$, let

$$v^{\alpha,\tau}(s,x) = E^{\alpha}_{s,x}[\int_0^{\tau} f^{\alpha_t}(s+t, x_t^{\alpha,s,x})e^{-\varphi_t}dt + H(s+\tau, x_\tau)e^{-\varphi_\tau}] \tag{2.24}$$

and

$$w(s,x) = \sup_{\alpha \in \mathcal{U}} \sup_{\tau \in \mathcal{M}(T-s)} v^{\alpha,\tau}(s,x). \tag{2.25}$$

Note that the E in (2.24) refers to taking expectation given x, s and α. For $s \leq T$, let

$$w_E(s,x) = \sup_{\alpha \in \mathcal{U}_E(s,x)} v^{\alpha,\tau}(s,x)$$

and

$$w_M(s,x) = \sup_{\alpha \in \mathcal{U}_M(s,x)} v^{\alpha,\tau}(s,x).$$

Theorem 9 *The function $w(s,x)$ is continuous with respect to (s,x) on $[0,T] \times \mathbf{R}^d$, $w(s,x) \geq H(s,x)$, $w(T,x) = H(T,x)$. Also, there exists a constant $N = N(m,k,T)$ such that for all $s \in [0,T]$, and $x \in \mathbf{R}^d$,*

$$\mid w(s,x) \mid \leq N(1+ \mid x \mid)^m.$$

□

Let $\varepsilon > 0$, and let

$$\tau_\varepsilon^{\alpha,s,x} = \inf(t \geq 0 : w(s+t, x_t^{\alpha,s,x})) \leq H(s+t, x_t^{\alpha,s,x}) + \varepsilon.$$

Note that $\tau_\varepsilon^{\alpha,s,x}$ is the time of the first exit of the process $(s+t, x_t^{\alpha,s,x})$ from the open set $Q_\varepsilon = \{(s,x) \mid w(s,x) > H(s,x) + \varepsilon\}$. It is clear that $\tau_\varepsilon^{\alpha,s,x} \in \mathcal{M}(T-s)$.

Theorem 10 $w(s,x) = w_E(s,x)$. *Furthermore, for* $s \in [0,T]$, $x \in \mathbf{R}^{\mathrm{d}}$, *the following inequality holds:*

$$w(s,x) \leq \sup_{\alpha \in \mathcal{U}_E(s,x)} E_{s,x}^{\alpha}\Big[\int_0^{\tau_\varepsilon} f^{\alpha_t}(s+t, x_t)e^{-\varphi_t}dt + H(s+\tau_\varepsilon, x_{\tau_\varepsilon})e^{-\varphi_{\tau_\varepsilon}}\Big] + \varepsilon.$$

□

This theorem shows that $\tau_\varepsilon^{\alpha,s,x}$ is an ε-optimal stopping time of the controlled process.

If A consists of a single point, the last inequality becomes an equality for $\varepsilon = 0$.

2. Friedman [15]

Let Q be a bounded domain in the $(n+1)$-dimensional space of variables (x,t). Assume that Q lies in the strip $0 < t \leq T$ and that $\widetilde{B} = \bar{Q} \cap \{t = 0\}$,$\widetilde{B}_T = \bar{Q} \cap \{t = T\}$ are nonempty. Let B_T be the interior of $\widetilde{B}_T$, B be the interior of $\widetilde{B}$. Denote by S_0 the boundary of Q lying in the strip $0 < t \leq T$, and let $S = S_0 \setminus B_T$. The set $\partial_0 Q = B \cup S$ is called the normal (or parabolic) boundary of Q. Let

$$\mathcal{L} = \frac{1}{2}\sum_{i,j=1}^{n} a_{i,j}(x,t)\frac{\partial^2}{\partial x_i \partial x_j} + \sum_{i=1}^{n} b_i(x,t)\frac{\partial}{\partial x_i} + c(x,t) - \frac{\partial}{\partial t}$$

be a uniformly parabolic operator in Q. Consider the initial boundary value problem of finding a solution u of

$$\begin{aligned}
&\mathcal{L}u(x,t) = f(x,t) \text{ in } Q \cup B_T \\
&u(x,0) = \varphi(x) \text{ on } B \\
&u(x,t) = g(x,t) \text{ on } S
\end{aligned}$$

where f, φ, g are given functions. If $g = \varphi$ on $\bar{B} \cap \bar{S}$, then the solution u is always understood to be continuous in $\bar{Q}$.

Theorem 11 *Assume that $\mathcal{L}$ is uniformly parabolic in Q, that a_{ij}, b_j, c, f are uniformly Hölder continuous in $\bar{Q}$, and that g,φ are continuous functions on $\bar{B}$ and $\bar{S}$ respectively, and $g = \varphi$ on $\bar{B} \cap \bar{S}$. Then there exists a unique solution u of the initial boundary value problem above.*

□

3. Jacka [23]

Let $(\xi_t)_{t \geq 0}$ be a diffusion in $\mathbf{R}^d$ (with respect to $(\mathcal{F}_t)$), and $X_t = e^{-\alpha t} g(\xi_t, T-t)$, where g is a continuous function. If

$$S_t = \text{ess sup}_{t \leq \tau} E\left[X_\tau \mid \mathcal{F}_t\right],$$

then clearly $S_t = e^{-\alpha t} f(\xi_t, T-t)$, where f is a continuous function and $f \geq g$ (Krylov [35]). Let $D \equiv \{(x,s) : f(x,s) > g(x,s)\}$. Because f and g are continuous, it is apparent that the local time $L^0(S-X)$ (see Protter [50], §4.5 for further details) will only increase for $\xi_t \in \partial D$.

Next, let $X = M + A$ be the canonical decomposition of X, where M is a martingale and A is a continuous, predictable process of integrable variation with $A_0 = 0$. Denote by A^- the decreasing component of A. Then we have the following theorem and corollary:

Theorem 12 *If ξ, f, g are defined as above, and there exist (deterministic) measures m_1, m_2 such that*

(i) ξ has density ρ with respect to m_1,

(ii) $dA^- \ll dm_2$,

(iii) $m_1 \otimes m_2(\partial D) = 0$,

then $L^0(S - X)$ is indistinguishable from 0.

PROOF. Let $\mu_t \equiv (1/2)dL_t^0/dA_t^-$. It follows from the strong Markov property that μ and $K \equiv dA_s^-/dm_2(s)$ are of the form $\mu_s = \mu(\xi_s, T-s)$ and $K_s = e^{-\alpha t}K(\xi_s, T-s)$, respectively. Denoting the t-section of ∂D by $(\partial D)_t$, we have

$$
\begin{aligned}
&EL^0(S-X)\\
&= E\int_0^t \mu_s \mathbf{1}_{(S-X)_s>0} dA_s^-\\
&= E\int_0^t \mu_s \mathbf{1}_{(\xi_s,T-s)\in\partial D} dA_s^-\\
&= \int_0^t \int_{\partial D_s} e^{-\alpha s}\mu(a,T-s)\rho(\xi_0,a;s)K(a,T-s)dm_1(a)dm_2(s)\\
&= \int_{\partial D\cap(\mathbf{R}^d\times[0,t])} e^{-\alpha s}\mu(a,T-s)\rho(\xi_0,a;s)K(a,T-s)d(m_1\otimes m_2)(a,s)\\
&= 0
\end{aligned}
$$

by (iii). □

REMARK. Under the conditions of the above theorem, we can represent f as follows

$$
\begin{aligned}
&f(\xi_0,T)\\
&\quad = A_0 + E\int_0^T \mathbf{1}_{(\xi_s,T-s)\in D} dA_s\\
&\quad = E\{A_T - \int_0^T \mathbf{1}_{(\xi_s,T-s)\in D^c} dA_s\}\\
&\quad = E\{e^{-\alpha T}g(S_T,0) + \int_0^T e^{-\alpha s}K(\xi_s,T-s)ds\}
\end{aligned}
$$

$$= \int_{\mathbf{R}^d} e^{-\alpha T} \rho(\xi_0, a; T) g(a, 0) dm_1(a)$$
$$+ \int_{D^c \cap (\mathbf{R}^d \times [0,T])} e^{-\alpha s} \rho(\xi_0, a; s) K(a, T-s) d(m_1 \otimes m_2)(a, s).$$

Corollary 3 *Suppose the conditions of the above theorem are satisfied. If ρ is C^1 in ξ with derivatives which are uniformly continuous in $\mathbf{R}^d \times [t_0, t_1]$ for any $0 < t_0 < t_1 < \infty$, then f' is C^1 in ξ for all $t > 0$.*

□

4. Jacka [22]

Let f be as in (2.3).

Lemma 3 *$f(x,t) > 0$ for all $x \geq 0$,$t > 0$.*

PROOF. Note that $g(x) > 0$ for all $x < X$. Then for $x \geq X$ and $t > 0$, we have

$$f(x,t) \geq (X/2)(E_x e^{-r\tau_{X/2}} \mathbf{1}_{\tau_{X/2}<t}) > 0$$

where $\tau_{X/2} = \inf\{s \geq 0 : S_s \leq X/2\} \wedge t$. □

Proposition 3 *The boundary b is decreasing in t and bounded above by X.*

PROOF. We know that $f(x,\cdot)$ is increasing for each x, and $f(\cdot,t)$ is decreasing for each t, by the proof of Proposition 2 , so that for any $t > 0$, $s > 0$, $\varepsilon > 0$

$$\begin{aligned} f(b(t)+\varepsilon, t+s) &\geq f(b(t)+\varepsilon, t) \\ &> g(b(t)+\varepsilon), \end{aligned}$$

since $b(t) + \varepsilon \in D_t$, and so $b(t) + \varepsilon \in D_{t+s}$. Thus $b(t+s) \leq b(t)$, so b is decreasing in t. Finally, note that g vanishes on $[X, \infty)$, whereas if $x \geq X$, $t > 0$, then by Lemma 3, $f(x,t) > 0$ so $b(t) < X$. □

Proposition 4 *The boundary b is continuous.*

PROOF. Both f and g are continuous , so since $D = \{(x,t) : f-g > 0\}$ is open, D^c is closed. If $t_n \uparrow t$, then $(b(t_n), t_n) \in D^c$ for all n. Thus $b(t^-) \leq b(t)$. This gives us left continuity of b, since b is decreasing in t. Recall that $\mathcal{L}f = 0$ in D by Lemma 2, and

$$\frac{\sigma^2 x^2}{2}\frac{\partial^2 f}{\partial x^2} \geq rf$$

in D, since $\partial f/\partial t \geq 0$ and $\partial f/\partial x \leq 0$. Thus by letting

$$D^n \equiv D \cap [0, X] \times [\frac{1}{n}, n],$$

we obtain

$$\inf_{(x,t)\in D^n} \frac{\sigma^2 x^2}{2}\frac{\partial^2 f}{\partial x^2} \geq \varepsilon_n > 0$$

for some ε_n, since $f > 0$ in the closure of D^n and f is continuous while the closure of D^n is compact. Given $t > 0$, take N such that $t \geq 1/N$. Then for any $x \in [b(t^+) + \eta, X]$, we can write

$$\begin{aligned} & f(x,s) - g(x,s) \\ = \;& \int_{b(s)}^{x}\int_{b(s)}^{y}\left[\frac{\partial^2 f}{\partial x^2}(u,s) - \frac{\partial^2 g}{\partial x^2}(u,s)\right] dudy \\ > \;& \frac{\eta^2}{\sigma^2 X}\varepsilon_n \end{aligned}$$

since $(\sigma^2 x^2/2)(\partial^2 g/\partial x^2)$ vanishes on $[0, X] \times \mathbf{R}^+$, and f, g agree on b up to first derivatives and b is decreasing.

Take a sequence $(s_n) \downarrow t$, and use the continuity of $f-g$ to deduce that

$$f(b(t^+) + \eta, t) - g(b(t^+) + \eta, t) \geq \frac{\eta^2}{\sigma^2 X}\varepsilon_n > 0.$$

Thus $(b(t^+) + \eta) \in D_t$ for all $\eta > 0$. This implies that $b(t^+) \geq b(t)$. Since b is also decreasing, it follows that b is right continuous. The result then follows. □

Proposition 4. The φ and [illegible] is continuous.

Proof. Both f and g are continuous [illegible] ... D [illegible] is open, [illegible] is closed. [illegible] then $(t_n, x_n) \in E$ for all [illegible] [illegible] $< \alpha$ [illegible] ... we [illegible] continuity of b, since [illegible] is decreasing [illegible] [illegible] that [illegible] $= 0$ [illegible] by Lemma 2, and

$$\frac{\partial \varphi(\cdot)}{\partial x}$$

in E, since $\partial f/\partial x$ [illegible] and ∂ [illegible] 0. Thus by letting

$$D' = E \cap ([0, X] \times [illegible])$$

we obtain

$$\inf_{x \in D'} \frac{\partial^2 \varphi}{\partial x^2} [illegible] > 0$$

for some [illegible], since $f > 0$ in the closure of D' and f is continuous while the closure of D' is compact. Given $\varepsilon > 0$, pick N such that [illegible] $\geq N$. Then for all $(t, x) \in [illegible] \in E$, we can write

$$[illegible] = g(x, s)$$

$$= \int [illegible] \int_{(t)} \left[\frac{\partial^2}{\partial x^2}(u, s) - \frac{\partial^2 g}{\partial x^2}(u, s) \right] du ds$$

$$\geq \frac{[illegible]}{[illegible]}$$

since [illegible] and [illegible] vanishes on $[0, X] \times \mathbb{R}^+$ and [illegible] agree on [illegible] to first derivatives and [illegible] is decreasing.

Take a sequence [illegible], and use the continuity of [illegible] to deduce that

$$[illegible] \; f(\varphi(t) + \eta, t) - g(b(t), t) [illegible] > 0.$$

Thus $(b(t)) + \eta) \notin E$ for all [illegible] > 0. This implies that $b(t)$ [illegible] $\geq b(t)$. Since b is also decreasing, it follows that b is right continuous. The remaining part follows. $\square$

Chapter 3
Analytic Approximation of the American Put

As stated earlier, Black and Scholes [5] derived the pricing equation for an European put option when the stock prices follow geometric Brownian motion. For the same model, Merton [44] derived the pricing equation for an American put with infinite time to maturity (see Chapter 1). In this chapter, several analytic approximations to the American put price will be reviewed. Numerical solutions, which will be discussed in the next chapter, are generally expensive and do not provide much insight. By studying analytic approximations, we hope to gain insight into the problem of pricing.

No one has yet been able to find the general solution of the American put price, P, although P is known for several special situations (Johnson, [27]):

- $rT = 0$, $P = p$, i.e. the American and European puts are equivalent.
- $\sigma^2 T = 0$, $P = (X - S)^+$, i.e. the put is either worthless or exercised immediately.
- $X = 0$, $P = 0$.
- $S = 0$, $P = X$.
- $S = \infty$, $P = 0$.
- $X = \infty$, $P = \infty$.

and Merton [44]:

- $T = \infty$,

$$P = \begin{cases} X/(1+M)(S_c/S)^M & \text{if } S \geq S_c = MX/(1+M) \\ X - S & \text{if } S \leq S_c \end{cases}$$

notation being as defined in Chapter 1. We write Φ for the distribution function of a standard normal random variable. Again, S_c, the critical stock price (i.e., the price below which early exercise occurs) is, in general, an unknown function of X, M, and $\sigma^2 T$.

In this chapter, the works of Johnson [27], Geske and Johnson [16], MacMillan [38], Barone-Adesi and Whaley [3], and Lamberton [37] will be reviewed.

3.1 Johnson

Using the fact that an American put is more valuable than a European put, but less valuable than a European put with an exercise price which is constant in present value terms, Johnson develops an approximate analytic expression for the price of an American put on a stock which is not paying dividends. The expression is very accurate for all values of the risk-free rate so far observed in the United States. More precisely, Johnson derives a general inequality by noting that an investor would be willing to trade an American put with exercise price X and time to maturity T for an American put with the same terms, but with a step-up of $X(e^{rT} - 1)$ at maturity. This latter put is equivalent to another European put (Margrabe [39]); hence, one can write

$$p(X) \leq P(X) \leq p(Xe^{rT}). \tag{3.1}$$

Note that any European put is equivalent to an American put with exercise price rising at the risk-free rate.

It might appear that inequality (3.1) provides an analytic expression for P, since we must have

$$P(X) = \alpha p(Xe^{rT}) + (1-\alpha)p(X) \tag{3.2}$$

with $0 \leq \alpha \leq 1$. Unfortunately, α is not a constant but a rather complicated function of S/X, rT, and $\sigma^2 T$. Johnson estimates a functional form of α by observing from Parkinson's tables [49] that α is between 0.2 and 0.25. A first attempt is made with a function of the form

$$\alpha = \frac{rT}{a_0 rT + a_1} \tag{3.3}$$

with $a_0 \approx 4$ and $a_1 \ll 1$, since then (3.3) gives values of α in the correct range, except when rT is small. Parkinson's tables were used to find α, by regressing rT/α on rT. This yields $a_0 = 3.9649$ and $a_1 = 0.03325$ with $R^2 = 0.9998$. For in-the-money puts, expression (3.3) gives values which are too large; the values are too small for out-of-the-money puts. Therefore, (3.3) was modified to permit α to become larger as S becomes smaller, while giving the same values as before when $S = X$. The following functional form was used:

$$\alpha = \left(\frac{rT}{a_0 rT + a_1}\right)^u \tag{3.4}$$

where

$$u = \frac{\log(S/S_c)}{\log(X/S_c)}. \tag{3.5}$$

Finding S_c , in general, requires the solution of

$$X - S_c = P(X, S_c, rT, \sigma^2 T). \tag{3.6}$$

Since the Amercian put pricing equation is not known, the equation for S_c must be estimated. For $\sigma^2 T = 0$, we must have $S_c = X$, while for

$\sigma^2 T = \infty$ (Merton, [44]), we must have $S_c = XM/(1+M)$. Thus, it seems natural to approximate the functional form for S_c as

$$S_c = X\left(\frac{M}{1+M}\right)^m \tag{3.7}$$

for some function m.

Note that for $r = 0$ there is no reason to exercise early, and, hence, S_c must be zero; (3.7) satisfies this requirement. In order to have $m = 0$ for $\sigma^2 T = 0$ and $m = 1$ for $\sigma^2 T = \infty$, one could try

$$m = \sigma^2 T/(b_0 \sigma^2 T + b_1) \tag{3.8}$$

where b_0 should be one. In order to estimate b_0 and b_1, rewrite (3.7) and (3.8) as

$$m = \frac{\sigma^2 T}{b_0 \sigma^2 T + b_1} = \frac{\log(S_c/X)}{\log(M/(1+M))} \tag{3.9}$$

and (3.5) as

$$u = \frac{\log((S/X)(X/S_c))}{\log(X/S_c)} = \frac{\log \alpha}{\log(rT/(a_0 rT + a_1))}. \tag{3.10}$$

Now, regress

$$\frac{\sigma^2 T}{\log(S/X)}\{1 - \log(\alpha - rT/(a_0 rT + a_1))\}\log(M/(1+M)) \tag{3.11}$$

on $\sigma^2 T$. Since no listed puts have ever had $T > 0.12r^{-1}$, cases of large rT were excluded from the regression. Again, values for α were obtained using Parkinson's tables; a_0 and a_1 were found in the previous regression. The second regression yielded $b_0 = 1.04083$ and $b_1 = 0.00963$, with $R^2 = .9975$.

The formula developed by Johnson is valid only for American puts on stocks which do not pay dividends. The approximation is not very accurate for puts on stocks which pay large dividends.

To find the hedge ratio (the number of shares of stock to options), rewrite (3.2) as

$$\begin{aligned} P &= \alpha[X\Phi(-d_{02}) - S\Phi(-d_{01})] \\ &\quad + (1-\alpha)[Xe^{-rT}\Phi(-d_2) - S\Phi(-d_1)] \end{aligned} \tag{3.12}$$

where d_1 and d_2 are as in (1.16),

$$d_{01} = \frac{\log(S/X) + \sigma^2 T/2}{\sigma\sqrt{T}} \tag{3.13}$$

$$d_{02} = d_{01} - \sigma\sqrt{T}. \tag{3.14}$$

In this representation, one can view the American put as a weighted average of two European puts, one using a zero interest rate and one using the true interest rate. From (3.12) one obtains

$$\begin{aligned} \frac{\partial P}{\partial S} &= -\alpha\Phi(-d_{01}) - (1-\alpha)\Phi(-d_1) \\ &\quad + \frac{\alpha \log(rT)/(a_0 rT + a_1)}{S\log(X/S_c)}[p(Xe^{rT}) - p(X)]. \end{aligned} \tag{3.15}$$

In general, of course, the derivative of an approximation formula is less accurate than the formula itself. Using (3.15) to set a hedge will not necessarily produce an entirely risk-free hedge.

3.2 Geske and Johnson

Geske and Johnson present an alternative analytic formula which allegedly satisfies the partial differential equation and boundary conditions that characterize the American put valuation problem. As usual, the Black-Scholes hypotheses of perfect markets, constant r and σ, no dividends, and geometric Brownian motion for the stock price are assumed. The current time is set at zero.

The key to their solution is the assumption that each exercise decision is a discrete event, whence a formula is derived as the discounted expected value of all future cash flows. The cash flows arise because the put can be exercised at one of the future times $t_1, t_2, \ldots$. The infinite series thus derived should be a continuous time solution of the free boundary value problem.

Since the assumption of geometric Brownian motion implies that the stock price at any future date is a lognormally distributed random variable, the correlation coefficient between the overlapping Brownian increments at close times t_1 and t_2 ($t_2 > t_1$) is given by $\rho_{12} = (t_1/t_2)^{1/2}$. At each instant, one will exercise the put if (a) the put has not already been exercised and (b) the payoff from exercising the put equals or exceeds the value of the put if it is not exercised. The critical stock price is independent of the current stock price and is determined from the free boundary whenever $X - S_c = P(S_c, T)$, for some $S = S_c$ and any T. Note that this can also be expressed as $\partial P/\partial S = -1$ when $S = S_c$. At the first instant, there is no probability that the put will already have been exercised, so one just integrates the exercise price less the future stock price at this date, and then discounts to the present. This yields two terms, one being the discounted exercise price times the probability that the stock price will be below $S_c^{(t_1)}$. At the next instant, one performs a similar integration up to $S_c^{(t_2)}$, the new critical stock price, but one must exclude all those cases where the put has been exercised at the first date. Again one obtains two terms, one being the discounted exercise price times the probability that the stock price at the first instant will be above the first critical stock price $S_c^{(t_1)}$, and the stock price at the second instant will be below the second critical stock price $S_c^{(t_2)}$, and so on. The correlation coefficient is negative between the argument for the last instant and the arguments of the previous ones, but positive between the arguments for the previous times. Intuitively, the put will be exercised at this instant if the stock price is below the critical stock price for this instant, given that it was not exercised at all previous instants because the stock price was always above the critical stock price. Proceeding in this way, one obtains the following formulae: Set

$$x_i = d_1(S_c^{(t_i)}, t_i), \tag{3.16}$$

$$y_i = d_2(S_c^{(t_i)}, t_i). \tag{3.17}$$

Then one has

$$P = Xw_2 - Sw_1 \tag{3.18}$$

where the weights w_1 and w_2 are given by

$$\begin{aligned} w_1 &= \Phi(-x_1) + \Phi(x_1, -x_2; -\rho_{12}) \\ &\quad + \Phi(x_1, x_2, -x_3; \rho_{12}, -\rho_{13}, -\rho_{13}) \\ &\quad + \cdots \end{aligned}$$

$$\begin{aligned} w_2 &= e^{-rt_1}\Phi(-y_1) + e^{-rt_2}\Phi(y_1, -y_2; -\rho_{12}) \\ &\quad + e^{-rt_3}\Phi(y_1, y_2, -y_3; \rho_{12}, -\rho_{13}, -\rho_{13}) \\ &\quad + \cdots \end{aligned}$$

and the correlation coefficients are $\rho_{ij} = i/\sqrt{j}$ for all $i, j \geq 1$. $\Phi(\cdot)$, $\Phi(\cdot, \cdot; \cdot)$, and $\Phi(\cdot, \cdot, \cdot; \cdot, \cdot, \cdot)$ are the standard cumulative univariate, bivariate, and trivariate normals, respectively. The series are both infinite.

Equation (3.18) cannot be used to compute actual numbers for the American put values, but it offers some insight into the portfolio which duplicates the American put payoffs. Note that if we differentiate this formula, we also gain some information regarding the sensitivity to all the specified parameters, and obtain some simplification in computing the hedge ratio. Since one assumes that the risk-free rate is known and constant, the portfolio of bonds represented by Xw_2 is equivalent to investing the same amount in a risk-free bond of whatever maturity. However, if one were to introduce uncertainty about future interest rates, then term structure effects could be important. Note that the formula implies that the duplicating portfolio for the out-of-the-money puts is skewed toward longer maturity bonds, while for the in-the-money puts it is skewed toward shorter maturities.

Recall that the European put option is simply a special case of an American put option with only one exercise boundary. Then, (3.18) reduces to the European put formula when $P(S,T) = (X - S)^+$ only holds at $T = 0$. The critical stock price is a time-dependent path of stock prices that separates the exercise from the no exercise region in such a way as to maximize the value of the American put. Just at the point where the stock price is equal to the critical price, the put value would decrease one dollar for a one dollar increase in the stock price (i.e., $\partial P/\partial S(S = S_c) = -1$). As $S \to S_c$, the sensitivity of the American put, $\partial P/\partial T$, tends to zero. Also, at this exercise point, the interest rate effect on the American put exactly offsets the variance effects.

Note that because the option is linearly homogeneous with respect to the stock price and exercise price, the existence of either partial derivative implies the other. Since the time to maturity only appears in the formula multiplied by the interest rate or variance rate, the existence of any two of the partial derivatives with respect to r, σ and T implies that of the third.

The partial derivative of the valuation formula (3.18)with respect to the stock price is the hedge ratio (the number of shares of stock to options) in the portfolio:

$$\frac{\partial P}{\partial S} = -w_1 < 0. \tag{3.19}$$

The negative sign indicates that as the stock price rises, the put price falls. The hedge ratio can be thought of as either the negative amount of stock (sold short) to which the put is equivalent, or under risk neutrality, as the discounted expected cash outflow (divided by the stock price) that the put holder will experience.

As the exercise price rises, the put value rises:

$$\frac{\partial P}{\partial X} = w_2 > 0. \tag{3.20}$$

This can be considered as the expected cash inflow (divided by the exercise price).

As the interest rate rises, the American put value falls:

$$\frac{\partial P}{\partial r} = -X dt[e^{-rt_1}\Phi(') + 2e^{-rt_2}\Phi(') + \ldots] \leq 0. \tag{3.21}$$

Since the present value of the bonds in the duplicating portfolio decreases as the interest rate increases. $\Phi(')$ denotes the derivative of Φ's parameters.

As the variance rate rises, the put value rises:

$$\frac{\partial P}{\partial \sigma^2} = X\frac{\sqrt{T}}{2\sigma}w_2' \geq 0. \tag{3.22}$$

An increase in volatility increases the probability of both high and low stock prices, and with the asymmetry of an option's contingent payoffs, increases the option value.

As the time to expiration increases, the American put value rises:

$$\frac{\partial P}{\partial T} = Xr[e^{-rt_1}\Phi(') + 2e^{-rt_2}\Phi(') + \ldots] + X\frac{\sigma}{2\sqrt{T}} \geq 0. \qquad (3.23)$$

The first four partial derivatives are functionally different from their corresponding European counterparts, but they do have the same sign and similar interpretation. The partial derivative with respect to time to expiration is strictly positive for the American put (provided $S > S_c$), while its sign is ambiguous for the European put. This ambiguity for the European put is obvious because more time helps if the put is out-of-the-money, but hurts if the holder wants to exercise immediately. The strictly positive sign for the American put is intuitively plausible because extending the life gives the holder more choices.

Geske and Johnson also show how to evaluate the American put formula (3.18) with a polynomial expression based upon an extrapolation from only a small number of exercise points to the infinite limit. The evaluation is very efficient because one is approximating an exact solution rather than the partial differential equation or the stock price process itself. Arbitrary accuracy can be obtained by adding exercise points. However, they showed that only a few (three) critical stock prices need to be computed in order to obtain penny accuracy. Unlike Brennan and Schwartz [6] or Parkinson [49], they do not approximate the partial differential equation, or the stock price process (Cox and Rubinstein, [9]). Instead, they evaluate their formula as an exact solution of the partial differential equation subject to the (discrete) free exercise boundary.

Let P_1 be the price of a put that can only be exercised at time T (i.e., at expiration); this option is just the European put, and we can write $P_1 = p$, the European put value. Let P_2 be the value of a put that can only be exercised at time $T/2$ or time T. Then

$$P_2 = Xe^{-rT/2}\Phi(-d_2(S_c^{(T/2)}, T/2) - S\Phi(-d_1(S_c^{(T/2)}, T/2))$$

$$+ Xe^{-rT}\Phi[d_2(S_c^{(T/2)}, T/2), -d_2(S_c^X, T); -1/\sqrt{2}]$$
$$- S\Phi[d_1(S_c^{(T/2)}, T/2), -d_1(X, T); -1/\sqrt{2}]. \qquad (3.24)$$

The critical stock price, $S_c^{(T/2)}$, solves

$$S = X - p(S, X, T/2, r, \sigma) = S_c^{(T/2)}. \qquad (3.25)$$

Similarly, let P_3 be the value of a put that can only be exercised at times $T/3$, $2T/3$, or T. Then

$$\begin{aligned} P_3 \quad = \quad & Xe^{-rT/3}\Phi[-d_2(S_c^{(T/3)}, T/3)] - S\Phi[-d_1(S_c^{(T/3)}, T/3)] \\ & + Xe^{-2rT/3}\Phi[d_2(S_c^{(T/3)}, T/3), -d_2(S_c^{(2T/3)}, 2T/3); -1/\sqrt{2}] \\ & - S\Phi[d_1(S_c^{(T/3)}, T/3), -d_1(S_c^{(2T/3)}, 2T/3); -1/\sqrt{2}] \\ & + Xe^{-rT}\Phi[d_1(S_c^{(T/3)}, T/3), -d_1(S_c^{(2T/3)}, 2T/3), -d_1(X, T); \\ & 1/\sqrt{2}, -1/\sqrt{3}, -2/\sqrt{3}] \\ & - S\Phi[d_2(S_c^{(T/3)}, T/3), d_2(S_c^{(2T/3)}, 2T/3), -d_2(X, T); \\ & 1/\sqrt{2}, -1/\sqrt{3}, -2/\sqrt{3}] \end{aligned} \qquad (3.26)$$

and the critical stock prices $S_c^{(T/3)}$ and $S_c^{(2T/3)}$ solve

$$S = X - P_2(S, X, 2T/3, r, \sigma) = S_c^{(T/3)} \qquad (3.27)$$

and

$$S = X - p(S, X, T/3, r, \sigma) = S_c^{(2T/3)} \qquad (3.28)$$

respectively.

The values P_1, P_2, P_3, $\cdots$ define a sequence whose limit should be the American put value. Many techniques are available for computing such limits. One method is Richardson extrapolation (Dahlquist and

Bjorck [11], p.269). In this case, the quantity to be determined is the American put price for a particular set of values S, X, T, r, and σ. The step length is the time between points at which exercise is permitted. The version of Richardson extrapolation (Appendix, equation (3.64)) leads to the following equation:

$$P = P_3 + 7/2(P_3 - P_2) - 1/2(P_2 - P_1). \tag{3.29}$$

This polynomial can be used to determine American put values and hedge ratios.

Preliminary evidence indicates that the analytic formula evaluation tabulated is faster to compute, by a factor of 10, than numerical methods. This is because the binomial and finite difference methods compute n critical stock prices ($n = 150$ in Cox and Rubinstein) while here only three points are used. (Even the tabulated four point method computes six.) Note that the three point extrapolation is about twice as fast as the four point.

Recall that the hedge ratio for the American put is given approximately by $\partial P/\partial S = -w_1$. In the absence of an analytic formula the hedge ratio is numerically approximated by computing two put values for two different stock prices and then using a difference equation to approximate the partial derivative at an intermediate stock price.

Geske and Johnson attribute their solution to an economic interpretation. First,the risk free hedge allows economists to avoid the transformations required for solutions to partial differential equations. Second, compound option theory provided a straightforward method for interpreting the infinite series of interrelated probability integrals arising from the free boundary condition. A key to the solution is that each exercise decision is considered as a discrete event. Thus, the formula derived is a continuous time solution of the partial differential equation instants. The formula adds to our intuition because it implies an exact duplicating portfolio for the American put, consisting of specific positions in discount bonds and stock sold short.

The evaluation of the formula is a separate problem. At first blush the American put formula might be considered intractable due to the infinite series of integrals. However, since the formula is exact in the limit, arbitrary accuracy can be obtained by extrapolating from a small sequence of terms to the actual solution containing an infinite series. This formula evaluation procedure leads to a polynomial expression similar to that used to evaluate the integral terms in the Black-Scholes European put option formula.

3.3 MacMillan

A third analytic approximation for the value of the American put option is given by MacMillan. The approximation is claimed to give accurate values and is easy to implement on a computer.

The value of the American put option satisfies the following partial differential equation (1.25)

$$\frac{1}{2}\sigma^2 S^2 \frac{\partial^2 P}{\partial S^2} + rS\frac{\partial P}{\partial S} - rP - \frac{\partial P}{\partial T} = 0,$$

subject to the boundary conditions

$$P(S,T) = (X - S)^+; \tag{3.30}$$

$$P(S,t) \geq (X - S)^+; \tag{3.31}$$

$$P(S,t) \leq X; \tag{3.32}$$

$$\lim_{S\to\infty} P(S,t) = 0, \tag{3.33}$$

where T is the expiration date. In addition, we require equilibrium option prices to be continuous functions. If discontinuities arise, for example as S varies, then arbitrageurs (individuals who engages in arbitrage opportunities) would seek to buy on the low side of the discontinuity and sell on the high side; the resulting market forces would work to eliminate the discontinuity, and so at equilibrium no discontinuities could survive.

Recall that S_c is the price at which the put should rationally be exercised, defined by $P(S_c, t) = X - S_c$. Equation (1.25) is only applicable for $S \geq S_c$. For $S \leq S_c$, $P(S,t) = X - S$.

MacMillan finds an approximate solution of Merton's differential equation and its boundary conditions, in the following form: write

$$P = p + e, \tag{3.34}$$

so e is the *early exercise premium.* Since the European put satisfies an equation similar to the American put, e satisfies

$$\frac{1}{2}\sigma^2 S^2 \frac{\partial^2 e}{\partial S^2} + rS\frac{\partial e}{\partial S} - re - \frac{\partial e}{\partial T} = 0, \tag{3.35}$$

However the boundary conditions on e are substantially simpler than those on either P or p, since both P and p satisfy the same boundary condition:

$$P(S,T) = p(S,T) = (X - S)^+. \tag{3.36}$$

We are thus left with the simpler problem of solving (3.35) subject to the conditions $e(S,T) = 0$, $e(S,t) \leq X$, and $\lim_{S\to\infty} e(S,t) = 0$. Replacing T by a function $K(T)$, transforms (3.35) into

$$S^2\frac{\partial^2 e}{\partial S^2} + MS\frac{\partial e}{\partial S} - Me - \frac{M}{r}\frac{dK}{dT}\frac{\partial e}{\partial K} = 0, \tag{3.37}$$

where $M = 2r/\sigma^2$. Now write $e(S,K)$ in the form:

$$e(S,K) = K(T)f(S,K). \tag{3.38}$$

One then has

$$\begin{aligned}
\frac{\partial e}{\partial S} &= K\frac{\partial f}{\partial S};\\
\frac{\partial^2 e}{\partial S^2} &= K\frac{\partial^2 f}{\partial S^2};\\
\frac{\partial e}{\partial T} &= \frac{dK}{dT}f + K\frac{dK}{dT}\frac{\partial f}{\partial K}.
\end{aligned}$$

Equation (3.35) then becomes

$$S^2K\frac{\partial^2 f}{\partial S^2} + MSK\frac{\partial f}{\partial S} - MKf - \frac{M}{r}[\frac{dK}{dT}f + K\frac{dK}{dT}\frac{\partial f}{\partial K}] = 0. \tag{3.39}$$

This can be rewritten as:

$$S^2\frac{\partial^2 f}{\partial S^2} + MS\frac{\partial f}{\partial S} - M[1 + \frac{dK/dT}{rK}(1 + K\frac{\partial f/\partial K}{f})]f = 0. \tag{3.40}$$

The function $K(T)$ is now chosen so that the time-dependence of $e = Kf$ is contained in the factor $K(T)$. A simple and useful choice is

$$K(T) = 1 - e^{-rT}. \tag{3.41}$$

(3.39) the becomes:

$$S^2\frac{\partial^2 f}{\partial S^2} + MS\frac{\partial f}{\partial S} - \frac{M}{K}[1 + (1-K)K\frac{\partial f/\partial K}{f}]f = 0.$$

The term $(1-K)K$ is zero at $T = 0$ (i.e. $K = 0$) and at $T = \infty$ (i.e. $K = 1$), and has a maximum value of $1/4$ at $K = 1/2$. Neglecting the term involving $(1-K)K$ should produce a useful approximation for small T and for large T, with error at intermediate values of T well controlled by the term $(1-K)K$, so consider

$$S^2\frac{\partial^2 f}{\partial S^2}+MS\frac{\partial f}{\partial S}-\frac{M}{K}f\approx 0. \tag{3.42}$$

One can assume that K is not equal to zero; the case $K=0$ is treated in the Appendix. The second-order ordinary differential equation (3.42) has two linearly independent solutions of the form aS^q. By substituting $f=aS^q$ into (3.42), one gets a quadratic equation in q:

$$q^2+(M-1)q-\frac{M}{K}=0. \tag{3.43}$$

This is easily solved for q to give two roots q_1 and q_2, with $q_1<0<q_2$, say. The general solution of (3.42) is

$$f(S)=a_1S^{q_1}+a_2S^{q_2}. \tag{3.44}$$

Necessarily $a_2=0$, since $S^{q_2}\to\infty$ with S, which is not allowed by condition (3.33). Writing $a=a_1$ and $q=q_1$, we obtain

$$P(S,T)=p+Kf=p(S,T)+(1-e^{-rT})aS^q.$$

Let $S=S_c$ be the value of S at which the curve $P=p+Kf$ touches the line $Y=X-S$, so

$$P(S_c)=p(S_c)+K(T)f(S_c)=X-S_c. \tag{3.45}$$

Note that at $S=S_c$, the slope of the curve $P=p+Kf$ must equal the slope of the line $Y=X-S$, which is -1. That is,

$$\frac{d}{dS}P(S_c)=\frac{d}{dS}p(S_c)+K(T)\frac{d}{dS}f(S_c)=\frac{d}{dS}(X-S_c)=-1, \tag{3.46}$$

where S_c in parentheses indicates evaluation at S_c after all other operations have been performed. Finally, recall that for $S \leq S_c$, one has $P = X - S$.

Use (3.45) and (3.46) to determine S_c and a. First of all note that

$$\frac{d}{dS}p(S) = \frac{d}{dS}c(S) - 1 = \Phi(d_1(S)) - 1$$

(see, Eq(45) in Smith [54]). Then from (3.45) and (3.46) obtain

$$\frac{Kf(S_c)}{K\frac{d}{dS}f(S_c)} = \frac{X - S_c - p(S_c)}{-\Phi(d_1(S_c))}.$$

But $f(S) = aS^q$, so that $\frac{d}{dS}f(S) = aqS^{q-1}$. Thus,

$$\frac{f(S_c)}{\frac{d}{dS}f(S_c)} = \frac{aS_c^q}{qaS_c^{q-1}} = \frac{S_c}{q}.$$

Hence

$$S_c = \frac{-q(X - p(S_c))}{\Phi(d_1(S_c)) - q}. \tag{3.47}$$

Thus S_c is the value of S which satisfies the equation $S = G(S)$, where

$$G(S) = \frac{-q(X - p(S))}{\Phi(d_1(S)) - q}.$$

This equation will not yield an explicit formula for S_c but can be solved to any desired degree of accuracy using standard iterative techniques. Having obtained S_c, use (3.46) to obtain

$$a = \frac{\Phi(d_1(S_c))S_c^{1-q}}{-qK}.$$

Now let

$$A = aKS_c^q = \frac{S_c\Phi(d_1(S_c))}{-q}. \tag{3.48}$$

Then an approximate solution of (1.25) subject to the boundary conditions (3.30)-(3.33) is

$$P(S,T) = p(S,T) + A(S/S_c)^q. \tag{3.49}$$

The limit $T = \infty$ is exact, and the solution in this case has been obtained by Merton [44]. For $T = \infty$, one has $K = 1 - e^{-rT} = 1$. Thus, the quadratic equation, $q^2 + (M-1)q - M = 0$, has solutions $q = 1$ and $q = -M$. Select the negative root $q = -M$ as before. Obtain S_c from the equation

$$S_c = \frac{-q(X - p(S_c))}{\Phi(d_1(S_c)) - q}.$$

At $T = \infty$, $p(S) = 0$ and $\Phi(d_1(S)) = 1$, so that this equation becomes $S_c = MX/(1+M)$, whence $A = S_c\Phi(d_1(S_c))/(-q) = X/(1+M)$, and so finally

$$P = A\left(\frac{S}{S_c}\right)^q = \frac{X}{1+M}\left(\frac{S}{S_c}\right)^{-M},$$

for $S \geq S_c$, while $P = X - S$ for $S \leq S_c$. These are precisely the results obtained by Merton.

3.4 Barone-Adesi and Whaley

Barone-Adesi and Whaley extended MacMillan's work to the theory of pricing commodity option contracts. A commodity option represents the right to buy or sell a specific commodity at a specified price within a specified period of time. The exact nature of the underlying commodity varies and may be anything from a precious metal such as gold or silver to a financial instrument such as a Treasury bond or a foreign currency. Usually the commodity option is labelled by the nature of the underlying commodity. For example, if the commodity option is written on a common stock, it is referred to as a stock option, and if the commodity option is written on a foreign currency, it is referred as a foreign currency option.

Consider a general commodity option-pricing model. The assumptions used in Barone-Adesi and Whaley's analysis are consistent with Black-Scholes and Merton. First, the short-term interest rate, r, and the cost of carry for the commodity, b, are assumed to be constant, proportional rates. For a non-dividend paying stock, the cost of carry is equal to the riskless rate of interest (i.e. $b = r$), but, for the most other commodities, this is not the case. For the traditional agricultural commodities such as grain and livestock, the cost of carry exceeds the riskless rate by the cost of storage, insurance, deterioration,etc.

A second common assumption in the option pricing literature is that the underlying commodity price change movements follows the stochastic differential equation,

$$\frac{dS}{S} = \mu dt + \sigma dW, \tag{3.50}$$

where μ is the expected instantaneous relative price change of the commodity, σ is the instantaneous standard deviation, and W is Brownian motion.

Finally, assuming a riskless hedge between the option and the underlying commodity may be formed, the partial differential equation governing the movements of the commodity option price V through time is

$$\frac{1}{2}\sigma^2 S^2 \frac{\partial^2 V}{\partial S^2} + bS\frac{\partial V}{\partial S} - rV - \frac{\partial V}{\partial T} = 0. \qquad (3.51)$$

This equation, which first appeared in Merton, is the heart of the commodity option pricing discussion contained herein. Note that, when the cost of carry rate b is equal to the riskles rate of interest, the above differential equation reduces to that of Black and Scholes. The non-dividend stock option pricing is a special case of the more general commodity option pricing problem.

Barone-Adesi and Whaley derive a quadratic approximation to the American put value. The key insight into this approximation is that, if the above partial differential equation applies to both the American and European options, it also applies to the early exercise premium of the American option. Recall that the early exercise premium is defined by (3.34).

Letting $M = 2r/\sigma^2$, and $N = 2b/\sigma^2$, the differential equation satisfied by e is

$$S^2\frac{\partial^2 e}{\partial S^2} + NS\frac{\partial e}{\partial S} - Me - \frac{M}{r}\frac{\partial e}{\partial T} = 0. \qquad (3.52)$$

Next, let $e(S,K) = K(T)f(S,K)$ as in (3.38). It follows that

$$\begin{aligned}\frac{\partial e}{\partial S} &= K\frac{\partial f}{\partial S};\\ \frac{\partial^2 e}{\partial S^2} &= K\frac{\partial^2 f}{\partial S^2};\\ \frac{\partial e}{\partial T} &= \frac{dK}{dT}f + K\frac{dK}{dT}\frac{\partial f}{\partial K}.\end{aligned}$$

Substituting the above partial derivatives into (3.52) yields

$$S^2\frac{\partial^2 f}{\partial S^2} + NS\frac{\partial f}{\partial S} - M[1 + \frac{dK/dT}{rK}(1 + K\frac{\partial f/\partial K}{f})]f = 0. \qquad (3.53)$$

Choose $K(T) = 1 - e^{-rT}$ as in (3.41), and substitute to obtain

$$S^2 \frac{\partial^2 f}{\partial S^2} + NS \frac{\partial f}{\partial S} - \frac{M}{K} f - (1 - K) M \frac{\partial f}{\partial K} = 0. \qquad (3.54)$$

Now, assume that the last term on the left-hand side is negligible. For commodity options with either very short or very long times to expiration, this assumption is reasonable since, as $T \to 0$ or $T \to \infty$, $\partial f / \partial K \to 0$, and the term $(1 - K) M (\partial f / \partial K)$ disappears. Therefore, one has

$$S^2 \frac{\partial^2 f}{\partial S^2} + NS \frac{\partial f}{\partial S} - \frac{M}{K} f \approx 0. \qquad (3.55)$$

The characteristic equation for (3.55) is

$$q^2 + (N - 1) q - \frac{M}{K} = 0, \qquad (3.56)$$

with a unique negative root q_1. As in MacMillan, the only feasible solution is $f(S) = a_1 S^{q_1}$, whence

$$P(S, T) = p(S, T) + K a_1 S^{q_1}. \qquad (3.57)$$

The value of a_1 is

$$a_1 = -\frac{[1 - e^{(b-r)T} \Phi(d_1(S_c))]}{K q_1 S_c^{q_1 - 1}}, \qquad (3.58)$$

where $-e^{(b-r)T} \Phi(d_1(S_c))$ is the partial derivative of $p(S_c, T)$ with respect to S_c, $a_1 > 0$ since $q_1 < 0$, and all other terms are positive (for the definition of d_1 see (1.16)). The critical commodity price S_c is determined from

$$X - S_c = p(S_c, T) - [1 - e^{(b-r)T} \Phi(d_1(S_c))] S_c / q_1. \qquad (3.59)$$

To find S_c, observe that $S_c(0) = X$, and

$$S_c(\infty) = \frac{X}{1-(1/q_1(\infty))}, \tag{3.60}$$

where $q_1(\infty) = [(1-N) - \sqrt{(N-1)^2+4M}]/2$. It is worthwhile to point out that when the cost of carry b is equal to the riskless rate of interest r, this result is exactly the same as Merton. In equation (3.59), S_c is a decreasing function of the time to expiration, and has the following approximate analytic expression

$$S_c \approx S_c(\infty) + [X - S_c(\infty)]e^{h_1}, \tag{3.61}$$

where $h_1 = (bT - 2\sigma\sqrt{T})X/(X - S_c(\infty))$.

With S_c known, the approximate value of an American put option (3.57) written on a commodity becomes

$$\begin{aligned} P(S,T) &= \; p(S,T) + A_1(S/S_c)^{q_1} \quad \text{when } S > S_c \\ P(S,T) &= \; X - S \quad \text{when } S \leq S_c, \end{aligned} \tag{3.62}$$

where $A_1 = -(S_c/q_1)[1 - e^{(b-r)T}\Phi(d_1(S_c))]$. Note that $A_1 > 0$ since $q_1 < 0$, $S_c > 0$, and $\Phi(d_1(S_c)) < e^{-bT}$.

3.5 Lamberton

For a probabilistic approach, consider an American put option on one share, with exercise price X and maturity T. Following Karatzas [28] and Myneni [48], write the fair price of this option at time $t \in [0,T]$ as

$$V_t = \text{ess sup}_{\tau\in\mathcal{T}_{t,T}} E(e^{-r(\tau-t)}(X - S_\tau)^+ \mid \mathcal{F}_t),$$

where $\mathcal{T}_{t,T}$ is the family of stopping times $\tau \in [t,T]$. Due to the Markov property of the model, we have $V_t = u(t, S_t)$, where

$$\begin{aligned} u(t,x) &= \sup_{\tau \in \mathcal{T}_{t,T}} E(e^{-r(\tau-t)}(X - xe^{(r-\sigma^2/2)(\tau-t)+\sigma(W_\tau - W_t)})^+) \\ &= \sup_{\tau \in \mathcal{T}_{0,T-t}} Ee^{-r\tau}(X - xe^{(r-\sigma^2/2)\tau+\sigma W_\tau})^+. \end{aligned}$$

For $t \in [0,T]$, define the *critical price* at time t as

$$S_c(t) = \sup\{x \geq 0 \mid u(t,x) = (X-x)^+\}.$$

The function $t \mapsto S_c(t)$ is C^∞ on the interval $[0,T)$ (see Van Moerbeke [55] and Friedman [15]), and $\lim_{t\to T} S_c(t) = X$. Since $t \mapsto u(t,x)$ is nonincreasing, $t \mapsto S_c(t)$ must be nondecreasing.

From the theory of optimal stopping, we know that

$$V_0 = \sup_{\tau \in \mathcal{T}_{0,T}} Ee^{-r\tau}(X - S_\tau)^+ = Ee^{-r\tau^*}(X - S_{\tau^*})^+$$

with $\tau^* = \inf\{t \in [0,T] \mid u(t,S_t) = (X - S_t)^+\}$ being the smallest optimal stopping time (El Karoui, [12], 72-238). In fact, the following result of Jaillet et al. [26] (Remark 3.12) shows that τ^* is the only optimal stopping time.

Lemma 4 *If τ is a stopping time in $\mathcal{T}_{0,T}$ and satisfies*

$$V_0 = Ee^{-r\tau}(X - S_\tau)^+$$

then $\tau = \tau^*$ almost surely.

□

Note that on $\{\tau^* < T\}$, we have $\tau^* = \inf\{t \in [0,T] \mid S_t \leq S_c(t)\}$ and that, if $S_0 \leq S_c(0)$, then $\tau^* = 0$ almost surely.

Lemma 5 *Assume $S_0 = x > S_c(0)$. Then, the support of the distribution of τ^* is equal to $[0, T]$.*

PROOF. Since $S_t = xe^{(r-\sigma^2/2)t+\sigma W_t}$, we have, with the convention that $\inf \emptyset = T$,

$$\begin{aligned}\tau^* &= \inf\{t \in [0,T] \mid \log(x) + (r - \sigma^2/2)t + \sigma W_t = \log S_c(t)\} \\ &= \inf\{t \in [0,T] \mid (r - \sigma^2/2)t - \log S_c(t) + \sigma W_t = -\log(x)\}.\end{aligned}$$

Now, for any $\varepsilon \in (0, T)$ the distribution of the process

$$(W_t + \frac{1}{\sigma}((r - \sigma^2/2)t - \log S_c(t)))_{0\leq t\leq T-\varepsilon}$$

is equivalent to that of the standard Brownian motion on the interval $[0, T - \varepsilon]$, since S_c is smooth on $[0, T)$. Hence, the result follows from the well-known properties of the hitting times of Brownian motion. □

A natural way to approximate the function u is to introduce a random walk approximation of the driving Brownian motion $(W_t)_{0\leq t\leq T}$. Let $(\varepsilon_k)_{k\geq 1}$ be a sequence of i.i.d. square-integrable random variables, with $E(\varepsilon_1) = 0$ and $E(\varepsilon_1^2) = T$, and let

$$W_t^{(n)} = \frac{\sum_{k=1}^{[nt/T]} \varepsilon_k}{\sqrt{n}}.$$

For $t = kT/n$, $k = 1, \ldots, n$, let

$$u^{(n)}(t, x) = \sup_{\tau \in \mathcal{T}_{0,T-t}^{(n)}} E(e^{-r\tau} X - xe^{\sigma W_\tau^{(n)} - \sigma^2/2\tau})$$

where $\mathcal{T}_{0,T-t}^{(n)}$ is the set of stopping times (with respect to the natural filtration of $W^{(n)}$) with values in

$$[0, T-t] \cap \{0, T/n, 2T/n, \ldots, (n-1)T/n, T\}.$$

From the definition of $u^{(n)}$, it is clear that $x \mapsto u^{(n)}(t,x)$ is a nonincreasing, convex function, satisfying $u^{(n)}(t,x) \geq (X-x)^+$.

We extend the definition of $u^{(n)}$ to all of $[0,T]$ by setting

$$u^{(n)}(t,x) = u^{(n)}([nt/T]T/n, x),$$

where $[a]$ denotes the integer part of a. With this definition, $u^{(n)}$ is a nonincreasing function of t, whose values may be computed (by dynamic programming), by solving the following recursive equations:

$$u^{(n)}(T,x) = g(x)$$

$$u^{(n)}\left(\frac{kT}{n}, x\right) = \max(g(x), e^{-rT/n} E u^{(n)}\left(\frac{(k+1)T}{n}, xe^{\frac{(r-\sigma^2/2)T}{n} + \frac{\sigma\varepsilon_1}{\sqrt{n}}})\right)$$

for all integers k between 0 and $n-1$.

(Note that when the ε_k are Bernoulli random variables, the algorithm we have just described is a variant of the Cox-Ross-Rubinstein method [10] for valuing American put options. The methods of Parkinson approach [49] and Geske and Johnson [16] are also relevant to this approach.)

The main properties of the approximate critical price associated with the $u^{(n)}$ are as follows.

Proposition 5 *Let $\rho_n = \{0, T/n, 2T/n, \ldots, (n-1)T/n\}$. For all large enough n and all $t \in \rho_n$, there is a number $s_n(t) \in [0, X)$ such that $u^{(n)}(t,x) = (X-x)$ for all $x \in [0, s_n(t)]$, and $u^{(n)}(t,x) > (X-x)$ for all $x \in (s_n(t), X]$.*
Moreover, letting $S_t^{(n)} = xe^{(r-\sigma^2/2)t + \sigma W_t^{(n)}}$, the stopping time defined by

$$\tau_n^* = \inf\{t \in \rho_n \mid S_t^{(n)} \leq s_n(t)\}$$

satisfies

$$u^{(n)}(0,x) = E(e^{-r\tau_n^*}X - xe^{\sigma W_{\tau_n^*}^{(n)} - \sigma^2/2\tau_n^*})^+.$$

PROOF. First observe that for $x \geq 0$,

$$u^{(n)}\left(\frac{(n-1)T}{n}, x\right) = \max(g(x), e^{-rT/n}E(X - xe^{(r-\sigma^2/2)T/n+\sigma(\varepsilon_1/\sqrt{n})})^+).$$

Therefore, for all $\alpha > 0$,

$$\begin{aligned} u^{(n)}((n-1)T/n, X) &= Xe^{-rT/n}E(1 - e^{(r-\sigma^2/2)T/n+\sigma\varepsilon_1/\sqrt{n}})^+ \\ &\geq Xe^{-rT/n}(1 - e^{(r-\sigma^2/2)T/n-\sigma\alpha/\sqrt{n}})\Pr(\varepsilon_1 < -\alpha) \\ &= Xe^{-rT/n}\Pr(\varepsilon_1 < -\alpha)\left(\frac{\sigma\alpha}{\sqrt{n}} + o(\frac{1}{\sqrt{n}})\right). \end{aligned}$$

Since $E(\varepsilon_1) = 0$ and $E(\varepsilon_1^2) = T$, one can find $\alpha > 0$ such that $P(\varepsilon_1 < -\alpha) > 0$, so that, for n large enough,

$$u^{(n)}((n-1)T/n, X) > 0.$$

Since $t \mapsto u^{(n)}(t, X)$ is nonincreasing, we have $u^{(n)}(t, X) > 0$ for all $t \in \rho_n$. The existence of $s_n(t)$ now follows from the convexity of the function $x \mapsto u^{(n)}(t, x)$.

To see that τ_n^* is an optimal stopping time, recall that the smallest optimal stopping time is given by

$$\overline{\tau}_n = \inf\{t \in \rho_n \mid u^{(n)}(t, S_t^{(n)}) = (X - S_t^{(n)})^+\}.$$

Obviously, $\tau_n^* \geq \overline{\tau}_n$. The inequality may be strict, since $u^{(n)}(t,x)$ may vanish for large values of x (especially when the distribution of ε_1 has finite support). But on $\{\tau_n^* > \overline{\tau}_n\}$, we have $(X - S_{\overline{\tau}_n}^{(n)})^+ = 0$, and therefore

$$e^{-r\tau_n^*}(X - S_{\tau_n^*}^{(n)})^+ \geq e^{-r\overline{\tau}_n}(X - S_{\overline{\tau}_n}^{(n)})^+,$$

which implies that τ_n^* is optimal too. □

Define s_n for all $t \in [0, T]$ by setting $s_n(t) = s_n([nt/T]T/n)$. The main result is the following.

Theorem 13 *The sequence of functions s_n converges uniformly to S_c as n tends to infinity.*

PROOF. Since T is arbitrary, it suffices to show uniform convergence on any interval $[\varepsilon, T)$ where $0 < \varepsilon < T$, because $u^{(n)}$ and u depend on T only through $T - t$. Next, since s is continuous and nondecreasing and each s_n is nondecreasing and bounded by $X = \lim_{t \to T} S_c(t)$, it suffices to prove pointwise convergence on $(0, T)$.

Let $t_0 \in (0, T)$. Then we know that

$$\lim_{n \to \infty} u^{(n)}(t_0, x) = u(t_0, x)$$

(cf Kushner [36], esp. Section 8.2). Thus

$$\limsup_{n \to \infty} s_n(t_0) \leq S_c(t_0).$$

Indeed, if $x < \limsup_{n \to \infty} s_n(t_0)$, then we have $u^{(n)}(t_0, x) = X - x$ for infinitely many values of n, whence $u(t_0, x) = X - x$.

In order to prove that $\liminf_{n \to \infty} s_n(t_0) \geq S_c(t_0)$, fix $x > S_c(0)$ and define τ_n^* as in Proposition 5. Since τ_n^* is optimal for the discretized problem, any weak limit of the law of $(\tau_n^*, W^{(n)})$ can be interpreted as a law of (τ, W), where W is standard Brownian motion with respect to some filtration $\mathcal{F}$ and τ is an $\mathcal{F}$-stopping time which maximizes $E(Xe^{-r\tau} - xe^{\sigma W_\tau - \sigma^2\tau/2})^+$. Using Lemma 4, we know that the optimal stopping time for the limit is unique (note that the proof of Lemma 4 does not use the fact the filtration is the natural filtration of W, but only the fact that W is an $\mathcal{F}$ -Brownian motion). Therefore, $(\tau_n^*, S^{(n)}_{\tau_n^*})$ converges in law to (τ^*, S_{τ^*}).

Now assume that $\liminf_{n\to\infty} s_n(t_0) < S_c(t_0)$, and take $\varepsilon > 0$ such that $\liminf_{n\to\infty} s_n(t_0) < S_c(t_0) - \varepsilon$. Since S_c is continuous, there exists $\eta > 0$ such that for all $t \in [t_0 - \eta, t_0 + \eta]$,

$$\liminf_{n\to\infty} s_n(t_0) < S_c(t) - \varepsilon.$$

Choose a subsequence $(s_{n_k}(t_0))_{k\geq 0}$ such that for all k

$$s_{n_k}(t_0) < S_c(t_0 - \eta) - \varepsilon. \tag{3.63}$$

Since s_{n_k} and S_c are nondecreasing, we have

$$s_{n_k}(t) < S_c(t) - \varepsilon$$

for all k and all $t \in (t_0 - \eta, t_0)$. By Lemma 5, we have $\Pr(\tau^* \in (t_0 - \eta, t_0)) > 0$ since $x > S_c(0)$. Let

$$A = \{(t, y) \in \mathbf{R}^2 \mid \mathrm{t_0 - \eta < t < t_0 \quad and \quad y > S_c(t) - \varepsilon}\}.$$

The set A is an open subset of $\mathbf{R}^2$, since S_c is continuous, and

$$\Pr((\tau^*, S_{\tau^*}) \in A) = \Pr(\tau^* \in (t_0 - \eta, t_0)),$$

since on $\{\tau^* < T\}$ we have $S_{\tau^*} = S_c(\tau^*)$. As $(\tau_n^*, S^{(n)}_{\tau_n^*})$ converges in law to (τ^*, S_{τ^*}), we have

$$\liminf_{n\to\infty} \Pr((\tau_n^*, S_{\tau_n^*}^{(n)}) \in A) \geq \Pr((\tau^*, S_{\tau^*}) \in A).$$

Therefore, for n large enough,

$$\Pr\left[\tau_n^* \in (t_0 - \eta, t_0), S_{\tau_n^*}^{(n)} > S_c(\tau_n^*) - \varepsilon\right] > 0.$$

But on $\{\tau_n^* < T\}$ we have $s_n(\tau_n^*) \geq S_{\tau_n^*}^{(n)}$, so the above inequality contradicts (3.63). □

3.6 Appendix

The method of Richardson extrapolation is taken from Dahlquist and Bjorck [13] (p. 269-273). It is used when one is interested in knowing the limiting value of a certain quantity as the step length approaches zero.

Let $F(h)$ be the value of the function of interest when a step size of h is used. We wish to find $F(0)$. Suppose $F(h)$ takes the form

$$F(h) = F(0) + a_1 h^p + a_2 h^r + O(h^s)$$

where $s > r > p$. Then we can also write

$$F(kh) = F(0) + a_1 (kh)^p + a_2 (kh)^r + O(h^s)$$

and

$$F(qh) = F(0) + a_1 (qh)^p + a_2 (qh)^r + O(h^s)$$

where $q > k > 1$. Substituting for a_1 and a_2 and solving for $F(0)$ yields

$$F(0) = F(h) + \frac{A}{C}[F(h) - F(kh)] - \frac{B}{C}[F(kh) - F(qh)] \qquad (3.64)$$

where

$$A = q^r - q^p + k^p - k^r$$

$$B = k^r - k^p$$

$$C = q^r(k^p - 1) - q^p(k^r - 1) + k^r - k^p.$$

Using $P_1 = F(qh)$, $P_2 = F(kh)$, and $P_3 = F(h)$, we have $q = 3$, $k = 3/2$. If we expand $F(h)$ in a Taylor series around $F(0)$ and drop terms of the third order and higher, we obtain $p = 1$ and $r = 2$. Substitution into (3.64) gives (3.29). There is some error in (3.29) from dropping the higher order terms.

The Case $K = 0$

We are concerned here with the case $K = 1 - e^{rT} = 0$, that is, $rT = 0$. We assume $T \neq 0$, so that we deal only with the case $r = 0$. In the quadratic equation (3.43), the ratio M/K is then undefined. To deal with this case, suppose first that $r \neq 0$, and put $U = M/2 = r/\sigma^2$ and $V = \sigma^2 T$. Then $UV = rT$. For small rT expand K as

$$K = 1 - e^{rT} = 1 - (1 - rT + (rT)^2 + \cdots) = rT - (rT)^2 + \cdots.$$

Thus for r near zero K is approximately equal to rT, and thus M/K is approximately $M/rT = 2/V$, which is independent of r. Thus, for small r, equation (3.42) becomes

$$S^2\frac{\partial^2 f}{\partial S^2} + MS\frac{\partial f}{\partial S} - \frac{2}{V}f = 0.$$

Substituting $f = aS^q$ gives the quadratic equation

$$q^2 + (M-1)q - \frac{2}{V} = 0.$$

The negative root must be selected. The case $r = 0$ is obtained by setting $M = 0$. Since $r = 0$ implies $K = 0$, the solution in this case is exact.

Chapter 4
Numerical Methods

The Black-Scholes partial differential equation is relevant for many significant valuation problems where no analytic solutions have been found. This has led to considerable research employing numerical methods to find approximate solutions.

Efficient markets imply the rapid reflection of information in asset prices. Thus, the arrival of new information often will be accompanied by price changes. If the underlying asset is assumed to follow a diffusion process, then price changes are continuous. Alternatively, if the underlying asset is assumed to follow a jump process, then price changes are discontinuous. In the diffusion case, information is thought to arrive in a smooth, continuous fashion, and price changes can have either a constant or a changing variance, but with either a normal or a lognormal distribution. By contrast, a jump process signifies that the information arrival is discontinuous, and that the price changes have a Poisson distribution. In practice, it is suspected that a combined diffusion-jump process is generating the data.

The no-arbitrage partial equilibrium conditions have been derived for the pure diffusion, pure jump, and combined processes, and some analytic solutions have been found for each case. However, in many complex but realistic problems, numerical methods, including finite differences (Brennan and Schwartz) or numerical integration (Parkinson) must be employed to approximate the value of the assets.

The constant-variance diffusion approach to asset price changes leads to the partial differential equation (1.25). Despite the presence of variable coefficients, the Black-Scholes equation remains parabolic (this is essentially due to limited liability), although the approximation equations and the stability conditions become more complex with S.

Fortunately, a change of variables (cf [44]) can transform (1.25) into the following equation:

$$a\frac{\partial u}{\partial t} = \frac{\partial}{\partial x}\left(k\frac{\partial u}{\partial x}\right) \tag{4.1}$$

where a and k are C^2 functions of x and t. This may be a non- linear equation if a and k are allowed to vary with u as well as x and t. To use numerical schemes of high accuracy, equation (4.1) is transformed from variable to constant coefficients by making the transformation

$$y = \int \frac{dx}{k(x)}, \tag{4.2}$$

whereupon (4.1) becomes

$$\frac{\partial u}{\partial t} = \frac{1}{ak}\frac{\partial^2 u}{\partial y^2}. \tag{4.3}$$

In terms of the original variable x, this transformation scheme in y will have unequal spacing of the mesh points. Brennan and Schwartz [6] and Mason [40] have used a version of this transformation by substituting $y - \log(S)$ into equation (4.1).

A discussion of the works of Parkinson [49], Jaillet, Lamberton, and Lepeyre [26], Lamberton [37], and Allegretto, Barone-Adesi, and Elliott [1] follows.

4.1 Parkinson

Assume that one has an option which can only be exercised at definite times t_m, $m = 1, \ldots, N$. In the limit $N \longrightarrow \infty$, we obtain the true American option. In the time interval $t_{N-1} \leq t \leq t_N$, solve for the value $w(S, t_{N-1})$ of the option using the following equation

$$w(y,t) = e^{-r(T-t)} E_q[w(y',T) \mid y, T-t] \tag{4.4}$$

where q is the probability distribution defined, in terms of the density function $p(y,t) = \exp(f(x)t)$ of the stock price, by

$$q(y'-y,t) = \frac{1}{\pi} \int_{-\infty}^{+\infty} e^{ik(y'-y)+t(f(k)+ik(f(i)-r))} dk. \tag{4.5}$$

$w(S, t_{N-1})$ is then used as the boundary condition when (4.4) is applied to the computation of $w(S, t_{N-2})$, and so on, until the solution is obtained.

However, the computation done in this way will yield some put values which are less than $(X-S)^+$. Such values actually occur for European puts, since the option cannot be exercised until expiration. But for an American put, they cannot be allowed. In general, for the above procedure to be satisfactory, we must set

$$P(S,t_m) = \max\left(P(S,T), e^{-r(t_{m+1}-t_m)} E_q[P(S, t_{m+1} \mid S, t_{m+1} - t_m)]\right). \tag{4.6}$$

American puts will be more valuable than European puts because of the maximum taken in (4.6), whose effect is propagated back in time with each successive stage of the calculation; the longer the time until expiration, the greater the difference between them.

The above calculation is easily performed with the discrete forms of the relevant equations. For example, suppose the stock distribution is approximately lognormal, so $f(k) \approx -ika - k^2\sigma^2/2$, and

$$q(y,t) \approx \frac{1}{\sqrt{2\pi\sigma^2 t}} e^{\frac{-(y-(r-\sigma^2/2)t)^2}{2\sigma^2 t}}. \tag{4.7}$$

Now proceed as follows. Define $y_m = mdy$ for all integer values of m, $b_m = (1-e^{-y_m})(-y_m)^+$, $t_j = T - jdt$ for $j \geq 0$, $P(m,j) = P(y_m, t_j)$ for $j \geq 1$, and $P(m,0) = b_m$. Then the discrete version of (4.6) is

$$P(n, j+1) = \max(b_n, \sum_{m=-\infty}^{+\infty} q(y_n - y_m, dt)P(m,j)). \qquad (4.8)$$

One may use an approximation to (4.7) for substitution into (4.8). Define $u = (2r/\sigma^2 - 1)dy$, $a' = 0.25u/(1 + \sqrt{1 + 0.25u^2})$, $a = a' + 0.5$, $dt = 2a(1-a)(dy)^2/\sigma^2$, and define $c(m)$ by

$$\begin{cases} 0 & m \leq -2 \\ e^{-rdt}(1-a)^2 & m = -1 \\ e^{-rdt}2a(1-a) & m = 0 \\ e^{-rdt}a^2 & m = 1 \\ 0 & m \geq 2. \end{cases}$$

Then we may use

$$P(m, j+1) = \max(b_m, c(-1)P(m-1,j) + c(0)P(m,j) + c(1)P(m+1,j)) \qquad (4.9)$$

for $j \geq 0$ and all integer m. In the limit $dy \longrightarrow 0$, (4.9) converges to an exact equation for an American put, because the binomial distribution given by $c(m)$ produces the normal distribution given by (4.7).

For the actual numerical solution of (4.9), $dy \approx 0.032$ was used and $P(m,j)$ was set to 1 (respectively 0) for $m < -158$ (respectively $m > 32$), to concentrate on a practical region of interest. The numerical check performed by setting $b_m = 0$ agreed with the Black-Scholes value for the European put to three decimal places.

As expected, as r increases, a put option decreases in value. Parkinson includes a numerical comparison between formula (4.9) and the advertised put prices in the New York Times. On average, the percentage difference $(predicted/actual) - 1$ is -25 ± 8. Naturally, one would expect the advertised prices to be higher than the true values.

From this numerical comparison, Parkinson also concludes that most of the advertised puts are significantly overpriced. In addition, the values computed with $r = 0$ are in much better agreement with the data than those with $r = 7\%$.

4.2 Jaillet, Lamberton, and Lapeyre

A detailed discussion of the Brennan and Schwartz algorithm for the valuation of the American put options (except for the logarithmic change of variable) is provided. Although the formulation of the boundary condition in Brennan and Schwartz's paper was mathematically incorrect, the algorithm is completely justified. The main advantage to this approach is that the techniques of variational inequalities provide an adequate framework for the study of numerical methods.

We begin by studying the pricing formulae of Bensoussan [4] and Karatzas [28] from the point of view of diffusion models. Let $(\Omega, \mathcal{F}, \mathbf{P})$ be a probability space and $(W_t)_{t\geq 0}$ be a standard Brownian motion with values in $\mathbf{R}^n$. Denote by $(\mathcal{F}_t)_{t\geq 0}$ the $\mathbf{P}$-completion of the natural filtration of $(W_t)_{t\geq 0}$.

Consider a financial market with n risky assets, with prices S_t^1, S_t^2, $\ldots, S_t^n$ at time t, and let Y_t be the n-dimensional vector with components $Y_t^j = \log(S_t^j)$ for $j = 1, \ldots, n$. We will assume that Y_t satisfies the following stochastic differential equation:

$$dY_t = \beta(t, Y_t)dt + \sigma(t, Y_t)dW_t \tag{4.10}$$

on a finite interval $[0, T]$, where T is the horizon (namely, the date of maturity of the option).

We impose the following conditions on β and σ, and on the interest rate:

(H1) $\beta(t,x)$ is a bounded C^1 function from $[0,T] \times \mathbf{R^n}$ into $\mathbf{R^n}$, with bounded derivatives.

(H2) $\sigma(t,x)$ is a bounded C^1 function from $[0,T] \times \mathbf{R^n}$ into the space of $n \times n$ matrices, with bounded derivatives. Also, σ admits bounded continuous second derivatives $\frac{\partial^2 \sigma_{i,j}}{\partial x_i \partial x_j}$, satisfying a Hölder condition in x, uniformly with respect to (t,x) in $[0,T] \times \mathbf{R^n}$.

(H3) The entries $a_{i,j}$ of $\sigma(t,x)\sigma^*(t,x)/2$ (where $*$ denotes transposition) satisfy the following coercivity property: $\exists \eta > 0$ such that $\forall (t,x) \in [0,T] \times \mathbf{R^n}$ and $\forall \xi \in \mathbf{R^n}$ we have

$$\sum_{1 \leq i,j \leq n} a_{i,j}(t,x)\xi_i\xi_j \geq \eta \sum_{i=1}^{n} \xi_i^2$$

(H4) The instantaneous rate of interest $r(t)$ is a C^1 function from $[0,T]$ into $[0,\infty)$.

REMARK. Condition (H3) can be interpreted in terms of the completeness of the market (Bensoussan [4]; Karatzas [30],[29]; Harrison and Pliska [20]). .

An American contingent claim is defined by an adapted process $(h(t))_{0 \leq t \leq T}$, where $h(t)$ is the payoff of the claim if exercised at time t. The typical example is an American put option: for an option on one unit of asset i, with exercise price X, one has $h(t) = (X - S_t^i)^+$. Note that, for simplicity, we do not consider contingent claims allowing a payoff rate per unit of time as in Karatzas [28]. Assume that $h(t)$ depends only on the prices of risky assets at time t, so that $h(t) = \psi(Y_t)$, where ψ is a continuous function from $\mathbf{R^n}$ into $\mathbf{R}$. Note that in the case of a put on asset i, one has $\psi(x_i) = (X - e^{x_i})^+$.

We will denote by $(Y_s^{t,x})_{s \geq t}$ a continuous version of the flow of the stochastic differential equation (4.10). Therefore, $(s,t,x) \mapsto Y_s^{t,x}(\omega)$ is continuous for almost all ω, $Y_t^{t,x} = x$, and $Y^{t,x}$ satisfies (4.10) on $[t,T]$.

Proposition 6 *Assume ψ is continuous and satisfies $|\psi(x)| \leq Me^{M|x|}$ for some $M > 0$, and define a function u^* on $[0,T] \times \mathbf{R}^n$, by*

$$u^*(t,x) = \sup_{\tau \in \mathcal{T}_{t,T}} E(e^{-\int_t^\tau r(s)ds} \psi(Y_\tau^{t,x})), \tag{4.11}$$

where $\mathcal{T}_{t,T}$ is the set of all stopping times with values in $[t,T]$. Then u^* is continuous and, for any solution (Y_t) of (4.10)

$$u^*(t,Y_t) = \text{ess sup}_{\tau \in \mathcal{T}_{t,T}} E\left(e^{-\int_t^\tau r(s)ds} \psi(Y_\tau) \mid \mathcal{F}_t\right). \tag{4.12}$$

Note that if, under probability $\mathbf{P}$, the discounted vector price is a martingale, then (4.12) means that $u^*(t,Y_t)$ is the 'fair price' of the American option defined by ψ at time t (cf. Karatzas [28], [30], [29]; Bensoussan [4]).

PROOF. Using the equality

$$Y_s^{t,x} = x + \int_t^s \beta(v, Y_v^{t,x})dv + \int_t^s \sigma(v, Y_v^{t,x})dW_v, \tag{4.13}$$

it is easy to prove that

$$E\left(\sup_{t \leq s \leq T} e^{M|Y_s^{t,x}|}\right) \leq Ce^{M|x|}, \tag{4.14}$$

where C depends only on T, M, and on the bounds on β , σ. Now, observe that if $(t_1,x_1), (t_2,x_2) \in [0,T] \times \mathbf{R}^n$, with $t_1 < t_2$, then

$$\begin{aligned} &u^*(t_2,x_2) - u^*(t_1,x_1) = \\ &\quad \sup_{\tau \in \mathcal{T}_{t_2,T}} E(e^{-\int_{t_2}^\tau r(s)ds} \psi(Y_\tau^{t_2,x_2})) - \sup_{\tau \in \mathcal{T}_{t_2,T}} E(e^{-\int_{t_1}^\tau r(s)ds} \psi(Y_\tau^{t_1,x_1})) \\ &\quad + \sup_{\tau \in \mathcal{T}_{t_2,T}} E(e^{-\int_{t_1}^\tau r(s)ds} \psi(Y_\tau^{t_1,x_1})) - \sup_{\tau \in \mathcal{T}_{t_1,T}} E(e^{-\int_{t_1}^\tau r(s)ds} \psi(Y_\tau^{t_1,x_1})). \end{aligned}$$

Therefore,

$$
\begin{aligned}
&| u^*(t_2, x_2) - u^*(t_1, x_1) | \\
&\quad \leq E(\sup_{t_2 \leq s \leq T} | e^{-\int_{t_2}^{s} r(v)dv} \psi(Y_s^{t_2,x_2}) - e^{-\int_{t_1}^{s} r(v)dv} \psi(Y_s^{t_1,x_1}) |) + \\
&\quad +E(\sup_{t_1 \leq s \leq t_2} | e^{-\int_{t_1}^{s} r(v)dv} \psi(Y_s^{t_1,x_1}) - e^{-\int_{t_1}^{s} r(v)dv} \psi(Y_{t_2}^{t_1,x_1}) |),
\end{aligned}
$$

and the continuity of u^* follows from the continuity of ψ and of the flow and from (4.14) (which, applied with $2M$ instead of M, ensures uniform integrability).

The equality (4.12) can be proved directly, by arguing that the supremum in (4.11) and the essential supremum in (4.12) are the same when $\mathcal{T}_{t,T}$ is replaced by the set of stopping times with respect to the filtration $(\mathcal{G}_{t,s})_{s\geq t}$ of the increments $W_s - W_t$ for $s \geq t$ (cf. [25] for details). □

Next, we turn to the algorithm of Brennan and Schwartz. The inner product of two vectors $u, v \in \mathbf{R}^n$ will be denoted by (u, v). Let $| u |^2 = (u, u)$, and write $u \geq v$ if $u_i \geq v_i$ for all i.

Proposition 7 *Let A be an $n \times n$ matrix and $u, \varphi, \theta \in \mathbf{R}^n$. The following two systems are equivalent:*

$$Au \geq \theta, u \geq \varphi, (Au - \theta, \varphi - u) = 0, \tag{4.15}$$

$$u \geq \varphi, (Au - \theta, v - u) = 0 \tag{4.16}$$

for all $v \geq \varphi$.

□

This proposition, which expresses the linear complementarity problem as a variational inequality is well known (cf Kinderlehrer and Stampacchia, [34]). In *loc cit*, the following is also proved.

Theorem 14 *The linear complementary problem* (4.16) *has a unique solution for all* θ, φ *if and only if* A *has positive principal minors.*

□

Note that this property is satisfied if A is *coercive* in the following sense: There exists $C > 0$ such that for all $x \in \mathbf{R}^n$

$$(Ax, x) \geq C \mid x \mid^2 .$$

The following is a characterization of the solution.

Proposition 8 *Assume the nondiagonal coefficients of a coercive matrix are nonpositive. Then the solution of* (4.15) *is the smallest vector* u *satisfying* $Au \geq \theta$ *and* $u \geq \varphi$.

□

In the next two propositions, we assume that A is a tridiagonal matrix of the form

$$\begin{pmatrix} b_1 & c_1 & 0 & \dots & 0 \\ a_2 & b_2 & c_2 & \dots & 0 \\ 0 & \vdots & \vdots & \vdots & 0 \\ \vdots & \vdots & a_{n-1} & b_{n-1} & c_{n-1} \\ 0 & \dots & 0 & a_n & b_n \end{pmatrix}.$$

Proposition 9 *If* A *is a tridiagonal matrix as above with* $c_i = 0$ *for all* $i \in \{1, \dots, n-1\}$, *and* $b_i > 0$ *for all* $i \in \{1, \dots, n\}$, *then, for all* $\theta, \varphi \in \mathbf{R}^n$, *the unique solution of the system* (4.15) *can be computed by solving the following recursive relations*

$$u_1 = (\theta_1 / b_1) \vee \varphi_1$$

$$u_j = [(\theta_j - a_j u_{j-1}) / b_j] \vee \varphi_j$$

for $2 \leq j \leq n$.

PROOF. This is an immediate consequence of the equalities

$$(Au)_1 = b_1 u_1$$

$$(Au)_j = b_j u_j + a_j u_{j-1}$$

for $2 \leq j \leq n$. □

Given any coercive tridiagonal matrix A as above, consider the following lower triangular matrix $\widetilde{A}$

$$\begin{pmatrix} \widetilde{b}_1 & 0 & 0 & \dots & 0 \\ a_2 & \widetilde{b}_2 & 0 & \dots & 0 \\ 0 & \vdots & \vdots & \vdots & 0 \\ \vdots & \vdots & \vdots & \vdots & 0 \\ 0 & \dots & 0 & a_n & \widetilde{b}_n \end{pmatrix}$$

where $\widetilde{b_n} = b_n$ and $\widetilde{b_i} = b_i - [c_i/\widetilde{b_{i+1}}]a_{i+1}$ for all $1 \leq i \leq n-1$.

Lemma 6 *If A is coercive then $\widetilde{b_i} > 0$ for all $1 \leq i \leq n$.*

PROOF. Note that, for $1 \leq i \leq n$, $\prod_{j=i}^{n} \widetilde{b_j}$ is the determinant of the matrix derived from A by deleting the first $(i-1)$ rows and columns. This matrix, being a submatrix of A, is coercive and therefore, its determinant is positive. Hence $\widetilde{b_i} > 0$ for all i. □

If $\theta \in \mathbf{R}^n$, denote by $\widetilde{\theta}$ the vector with coordinates $\widetilde{\theta}_n = \theta_n$ and $\widetilde{\theta}_i = \theta_i - c_i\widetilde{\theta}_{i+1}/\widetilde{b}_{i+1}$ for $i < n$. Clearly, the two systems $Ax = \theta$ and $\widetilde{A}x = \widetilde{\theta}$ are equivalent.

Proposition 10 *Assume that A is a coercive tridiagonal matrix with $c_j \leq 0$ for all $1 \leq j \leq n-1$. If the solution u of system (4.15) satisfies $u_i = \varphi_i$ whenever $1 \leq i \leq k$, and $u_i > \varphi_i$ whenever $k < i$ for some $1 \leq k \leq n$, then u is also a solution of the following system:*

$$\widetilde{A}u \geq \widetilde{\theta}, u \geq \varphi, (\widetilde{A}u - \widetilde{\theta}, \varphi - u) = 0. \tag{4.17}$$

PROOF. Since $c_{i-1} \leq 0$ and $\widetilde{b}_i > 0$ (by Lemma 6), we have $-c_{i-1}/\widetilde{b}_i \geq 0$ for all $i \geq 2$. It follows that, for all $i \leq n-1$, there exist nonnegative numbers $\lambda_{i,i+1}, \ldots, \lambda_{i,n}$ such that

$$\widetilde{\theta}_i = \theta_i + \lambda_{i,i+1}\theta_{i+1} + \ldots + \lambda_{i,n}\theta_n$$

and

$$(\widetilde{A}u)_i = (Au)_i + \lambda_{i,i+1}(Au)_{i+1} + \ldots + \lambda_{i,n}(Au)_n.$$

Therefore, if u solves (4.15), we have $(\widetilde{A}u)_i \geq \widetilde{\theta}_i$. Moreover, by assumption, equality holds whenever $i > k$, since $(Au)_j$ is then equal to θ_j for all $j \geq i$. Hence, u solves (4.17). □

THE VALIDITY OF THE BRENNAN-SCHWARTZ ALGORITHM

Recall that for the classical models, the calculation of the price of the American put option reduces to solving the following system, in which we have made the change of variable $t \mapsto (T-t)$:

$$u \geq \psi,$$

$$\frac{\partial u}{\partial t} - \frac{\sigma^2}{2}\frac{\partial^2 u}{\partial x^2} - \beta\frac{\partial u}{\partial x} + ru \geq 0,$$

$$\left(\frac{\partial u}{\partial t} - \frac{\sigma^2}{2}\frac{\partial^2 u}{\partial x^2} - \beta\frac{\partial u}{\partial x} + ru\right)(\psi - u) = 0,$$

$$u(0) = \psi,$$

with $\psi(x) = (X - e^x)^+$. The implicit discretization scheme approximates the vector

$$(u(i\Delta t, j\Delta x))_{0 \leq i \leq N, n_1 \leq j \leq n_2}$$

by

$$(\bar{u}_j^i)_{0 \leq i \leq N, n_1 \leq j \leq n_2}$$

obtained by solving the following set of equations:

$$\bar{u}_j^0 = \psi(j\Delta x) \tag{4.18}$$

for $n_1 + 1 \leq j \leq n_2 + 1$,

$$a\bar{u}_{j-1}^i + b\bar{u}_j^i + c\bar{u}_{j+1}^i \geq \bar{u}_j^{i-1} \tag{4.19}$$

$$\bar{u}_j^i \geq \psi(j\Delta x) \tag{4.20}$$

$$(a\bar{u}_{j-1}^i + b\bar{u}_j^i + c\bar{u}_{j+1}^i - \bar{u}_j^{i-1})(\bar{u}_j^i - \psi(j\Delta x)) = 0 \tag{4.21}$$

$$\bar{u}_{n_1}^i = \psi(n_1\Delta x) \tag{4.22}$$

$$\bar{u}_{n_2}^i = 0 \tag{4.23}$$

for $1 \leq i \leq N$ and $n_1 + 1 \leq j \leq n_2 + 1$, where

$$a = -\frac{\Delta t \sigma^2}{2(\Delta x)^2} + \beta\frac{\Delta t}{2\Delta x},$$

$$b = 1 + \frac{\Delta t \sigma^2}{(\Delta x)^2} + r\Delta t,$$

$$c = -\frac{\Delta t \sigma^2}{2(\Delta x)^2} - \beta\frac{\Delta t}{2\Delta x},$$

and $N\Delta t = T$, with $[n_1\Delta x, n_2\Delta x]$ being a sufficiently large interval about $\log(X)$.

Note that (4.22) and (4.23) correspond to Dirichlet boundary conditions, whereas Brennan and Schwartz use mixed conditions, Dirichlet in $n_1\Delta x$ and Neumann in $n_2\Delta x$. The Brennan-Schwartz algorithm, as applied to the system (4.18)-(4.23), goes as follows:

(1) From (4.18) get $\bar{u}_j^0$ for $n_1 \leq j \leq n_2$.

(2) For $n_1 \leq j \leq n_2$, derive $(\bar{u}_j^i)$ from $(\bar{u}_j^{i-1})$ from the following equations:

$$\bar{u}_{n_1}^i = \psi(n_1\Delta x),$$

$$(\bar{u}_j^i) = \left(\frac{1}{\tilde{b}_j}(\tilde{u}_j^{i-1} - a\bar{u}_{j-1}^{i-1})\right) \vee \psi(j\Delta x),$$

for $n_1 + 1 \leq j \leq n_2 - 1$,

$$\bar{u}_{n_2}^i = \psi(n_2\Delta x) = 0$$

where

$$\tilde{b}_{n_2-1} = b,$$

$$\tilde{b}_{j-1} = b_{j-1} - ca/\tilde{b}_j,$$

$$\tilde{u}_{n_2-1}^{i-1} = \bar{u}_{n_2-1}^{i-1}$$

$$\tilde{u}_{j-1}^{i-1} = \bar{u}_{j-1}^{i-1} - c\tilde{u}_j^{i-1}/\tilde{b}_j,$$

for $n_1 + 2 \leq j \leq n_2 - 1$.

To justify the algorithm, first restate (4.18)-(4.23) as a set of variational inequalities. Let $n = n_2 - n_1 - 1$, and let u^i be the vector in $\mathbf{R}^n$ with components $u_j^i = \bar{u}_{j+n_1}^i$ for all j. Given any vector $v \in \mathbf{R}^n$, denote by $\dot{v}$ the vector with components $\dot{v}_1 = v_1 - a\psi(n_1\Delta x)$ and $\dot{v}_j = v_j$ for $j > 1$, and let A be the following $n \times n$ matrix:

$$\begin{pmatrix} b & c & 0 & \dots & 0 \\ a & b & c & \dots & 0 \\ 0 & \vdots & \vdots & \vdots & 0 \\ \vdots & \vdots & \vdots & \vdots & c \\ 0 & \dots & 0 & a & b \end{pmatrix}.$$

The system (4.18)-(4.23) is equivalent to the following set of variational inequalities:

$$\bar{u}_{n_1}^i = \psi(n_1\Delta x)$$

$$\bar{u}_{n_2}^i = \psi(n_2\Delta x)$$

for $0 \le i \le N$,

$$u^0 = \varphi$$

$$Au^i \ge \dot{u}^{i-1}$$

$$u^i \ge \varphi$$

$$(Au^i - \dot{u}^{i-1}, u^i - \varphi) = 0,$$

for $1 \le i \le N$, where φ is the vector with components $\varphi_j = \psi((n_1 + j)\Delta x)$ for all j.

It is clear that in order to justify the Brennan-Schwartz algorithm, it is sufficient to prove that the assumptions of Proposition 10 hold for the above variational inequalities. To this end, let us suppose that

$$\beta < \sigma^2/\Delta x. \tag{4.24}$$

The above, for example, holds when Δx is sufficiently small, and it implies that $a < 0$.

Lemma 7 *If* (4.24) *is satisfied, then A is a coercive matrix with the following property: if $x \neq 0$ and $Ax \ge 0$, then $x_i > 0$ for all i.*

PROOF. Using the fact that a and c are negative, we have

$$\begin{aligned}
(Ax, x) &= \sum_{j=2}^{n} a x_{j-1} x_j + \sum_{j=1}^{n} b x_j^2 + \sum_{j=1}^{n-1} c x_j x_{j+1} \\
&\ge \frac{a}{2} \sum_{j=2}^{n} (x_{j-1}^2 + x_j^2) + b \sum_{j=1}^{n} x_j^2 + \frac{c}{2} \sum_{j=1}^{n-1} (x_j^2 + x_{j+1}^2) \\
&\ge (a + b + c) \sum_{j=1}^{n} x_j^2 = (1 + r\Delta t) \sum_{j=1}^{n} x_j^2.
\end{aligned}$$

This proves coercivity. Next, suppose $Ax \ge 0$. Then $\tilde{A}x \ge 0$. Indeed, the numbers $\lambda_{i,j}$ introduced in the proof of Proposition 10 are positive since $a < 0$ and $c < 0$. If, in addition, $x \neq 0$, then $(\tilde{A}x)_1 > 0$, and using the negativity of a again, we find that all the components of x are positive. □

Lemma 8 *Let $\Gamma(\psi)$ be the set of all integers j such that*

$$a\psi((j-1)\Delta x) + b\psi(j\Delta x) + c\psi((j+1)\Delta x) \geq \psi(j\Delta x) > 0.$$

Then, $j \leq j_0 \in \Gamma(\psi)$ implies that $j \in \Gamma(\psi)$.

PROOF. Let $f(x) = X - e^x$ (so $\psi = f^+$). Using $c < 0$, it is easy to see that $\Gamma(\psi) \subseteq \Gamma(f)$. On the other hand, $\{j \mid \psi((j+1)\Delta x) \neq 0\} \subseteq \Gamma(\psi)$. Now for any integer j we have

$$\begin{aligned} &af((j-1)\Delta x) + bf(j\Delta x) + cf((j+1)\Delta x) - f(j\Delta x) \\ &= (a+b+c-1)X - e^{j\Delta x}(ae^{-\Delta x} + b + ce^{\Delta x} - 1). \end{aligned}$$

In the above equation, the right-hand side is a monotone function of j, whose limit, as j goes to $-\infty$, is $(a+b+c-1)X = Xrh \geq 0$. It follows that $j \leq j_0 \in \Gamma(f)$ implies that $j \in \Gamma(f)$, and the result follows. □

The above results, together with the following proposition, complete the justification of the Brennan-Schwartz algorithm.

Proposition 11 *Let $u^0, u^1, \ldots, u^N$ be the solution of the above variational inequalities. Then the following hold:*
1. $u^i \geq u^{i-1}$ for all $i \in \{1, \ldots, N\}$.
2. For each i, there exists k_i such that $u^i_j = \varphi_j$ for all $j \leq k_i$, and $u^i_j > \varphi_j$ for all $j > k_i$.

PROOF. Clearly, $u^1 \geq u^0 = \varphi$. If $u^i \geq u^{i-1}$, then $Au^{i+1} \geq u^{i-1}$, and $u^{i+1} \geq \varphi$, whence $u^{i+1} \geq u^i$ by Proposition 8.
As for (2), suppose first that $i = 1$. Since $\varphi \neq 0$, it follows from the previous lemma that $u^1_j > 0$ for all j. Suppose there exist integers ℓ_1, ℓ_2

with $1 \leq \ell_1 \leq \ell_2 \leq n$ such that $u^1_{\ell_2} = \varphi_{\ell_2}$, and $u^1_j > \varphi_j$ for all integers $j \in [\ell_1, \ell_2 - 1]$. By reducing ℓ_1 we may assume, if necessary, that either $\ell_1 = 1$ or $u_{\ell_1 - 1} = \varphi_{\ell_1 - 1}$. From $(Au^1)_{\ell_2} \geq \varphi_{\ell_2}$ and $u^1_{\ell_2} = \varphi_{\ell_2}$, it follows that

$$au^1_{\ell_2-1} + b\varphi_{\ell_2} + cu^1_{\ell_2+1} \geq \varphi_{\ell_2},$$

and so, since a and c are negative,

$$a\varphi_{\ell_2-1} + b\varphi_{\ell_2} + c\varphi_{\ell_2+1} \geq \varphi_{\ell_2}.$$

Therefore, $n_1 + \ell_2 \in \Gamma(\varphi)$, and so by the previous lemma again, $[n_1 + 1, n_1 + \ell_2] \subset \Gamma(\varphi)$, whence $(A\varphi)_j \geq \varphi_j$ for $j \in [1, \ell_2]$. Since for all j in $[\ell_1, \ell_2)$, one has $(Au^1)_j = \varphi_j$, it follows that u^1 satisfies $(Au^1)_j \leq (A\varphi)_j$ for all j in $[\ell_1, \ell_2)$. Therefore, $v = \varphi - u^1$ satisfies the conditions $(Av)_j \geq 0$ for $\ell_1 \leq j < \ell_2$, $v_{\ell_2} = 0$, and $v_{\ell_1} = 0$ if $\ell_1 > 1$. Applying Lemma 7 to a suitable submatrix of A, we find that $v_j \geq 0$ for $\ell_1 \leq j < \ell_2$. Hence, $u^1_j \leq \varphi_j$ for the same range of values of j. But this is a contradiction, so (2) holds for $i = 1$.

Finally, if (2) is true for some i, and $[\ell_1, \ell_2]$ is chosen such that $u^{i+1}_{\ell_2} = \varphi_{\ell_2}$ and $u^{i+1}_j > \varphi_j$ for all $j \in [\ell_1, \ell_2)$, then, using (1), we obtain $u^i_{\ell_2} = \varphi_{\ell_2}$, and so $u^i_j = \varphi_j$ for all $j \leq \ell_2$, and the same reasoning as above leads to a contradiction. □

REMARK. In the case of call options, a modified version of the Brennan-Schwartz algorithm can be justified, whereas the algorithm itself (as presented above) leads to a false solution. (The property to be used is $j \geq j_0 \in \Gamma(\psi)$ implies that $j \in \Gamma(\psi)$.)

For some functions ψ, such as $\psi(x) = |e^x - X|$, the Brennan-Schwartz algorithm fails to solve the variational inequality.

4.3 Lamberton

Lamberton's aim here is to prove the convergence of the critical price in the finite difference method of Brennan and Schwartz.

For $x \in \mathbf{R}$, let $\psi(x) = (X - e^x)^+$ and $\nu(t,x) = u(t, e^x)$ for $t \in [0,T]$. For $(t,x) \in [0,T] \times \mathbf{R}$, introduce the stochastic processes $(Y_s^{t,x})_{t \leq s \leq T}$, defined for $s \in [t,T]$, by

$$Y_s^{t,x} = x + (r - \sigma^2/2)(s-t) + \sigma(W_s - W_t)$$

where W is Brownian motion. With these notations, we have

$$\nu(t,x) = u(t,e^x) = \sup_{\tau \in \mathcal{T}_{t,T}} E(e^{-r(\tau - t)}\psi(Y_\tau^{t,x})),$$

and it follows from the definition of $S_c(t)$ that for all $t \in [0,T)$, we have

$$\log(S_c(t)) = \sup\{x \in \mathbf{R} \mid \nu(\mathrm{t},\mathrm{x}) = \psi(\mathrm{x})\}.$$

Before applying the finite difference method, we need to localize the problem in a bounded interval (see Jaillet et al [26], Section 4). Let M be a positive real number such that $e^{-M} < X < e^M$. For $(t,x) \in [0,T] \times \mathbf{R}$, define

$$T_M^{t,x} = \inf\{s \in [t,T] \mid \mid Y_s^{t,x} \mid \geq M\},$$

with the convention $\inf \emptyset = T$, and define

$$\nu_M(t,x) = \sup_{\tau \in \mathcal{T}_{t,T}} E(e^{-r(\tau \wedge T_M^{t,x} - t)}\psi(Y_{\tau \wedge T_M^{t,x}}^{t,x})).$$

Obviously, we have

$$\psi(x) \leq \nu_M(t,x) \leq \nu(t,x)$$

and

$$\lim_{M \to \infty} \nu_M(t,x) = \nu(t,x).$$

Proposition 12 *1. If $(t,x) \in [0,T) \times (-M,M)$, then $\nu_M(t,x) > 0$.*

2. For $t \in [0,T)$ define $s_M(t)$ by

$$\log(s_M(t)) = \sup\{x \in [-M,M) \mid \nu_M(t,x) = \psi(x)\}.$$

Then, $\log(s_M(t)) \in [-M, \log(X))$, and, $\forall x \in [-M, \log(s_M(t))]$, we have $\nu_M(t,x) = \psi(x)$.

3. The function $t \mapsto s_M(t)$ is continuous and nondecreasing on $[0,T)$ and, s_M converges uniformly to S_c as M tends to ∞.

PROOF. To prove (1), observe that

$$\nu_M(t,x) \geq E(e^{-r(T_M^{t,x}-t)}\psi(Y_{T_M^{t,x}}^{t,x})),$$

and that $\Pr(Y_{T_M^{t,x}}^{t,x} = -M) > 0$ if $x \in (-M,M)$, and $\psi(-M) = X - e^{-M} > 0$.

In order to prove (2), let $t \in [0,T)$, and

$$I_t = \{x \in [-M,M) \mid \nu_M(t,x) = \psi(x)\}.$$

It follows from (1) that $I_t \subset [-M, \log(X))$, since $\psi(x) = 0$ for $x \geq \log(X)$ and (2) will follow if we can prove that I_t is an interval. Now, let $x_1 < x_2$ belong to I_t. If $x \in [x_1, x_2]$, and we let

$$\rho_M^{t,x} = \inf\{s \in [t,T] \mid \nu_M(s, Y_s^{t,x}) = \psi(Y_s^{t,x})\},$$

then

$$\nu_M(t,x) = E(e^{-r(\rho_M^{t,x}-t)}\psi(Y_{\rho_M^{t,x}}^{t,x})).$$

By our choice of x, x_1, and x_2, we have $Y_s^{t,x} \in [x_1, x_2]$ on the set $\{s < \rho_M^{t,x}\}$, and therefore $\psi(Y_s^{t,x}) = X - e^{Y_s^{t,x}}$. But it is easy to check that the process $(e^{-r(s-t)}(X - e^{Y_s^{t,x}}))_{t \leq s \leq T}$ is a supermartingale. Hence $\nu_M(t,x) = \psi(x)$ and so $x \in I_t$.

The fact that s_M is nondecreasing follows from the fact that $t \mapsto \nu_M(t,x)$ is nonincreasing. The continuity of s_M can be proved by the same argument that establishes the continuity of S_c (cf Friedman [15], proof of Theorem 3.1). The convergence of s_M to S_c follows easily from the fact that $\nu_M(t,x) \uparrow \nu(t,x)$ as $M \uparrow \infty$. □

Now let $H_M = L^2(-M,M)$ and $V_M = H^1(-M,M) = \{f \in H_M \mid f' \in H_M\}$. For $f,g \in V_M$, let

$$\begin{aligned} a_M(f,g) &= \frac{\sigma^2}{2}\int_{-M}^{M} f'(x)g'(x)dx \\ &\quad - (r-\sigma^2/2)\int_{-M}^{M} f'(x)g(x)dx \\ &\quad + r\int_{-M}^{M} f(x)g(x)dx. \end{aligned}$$

Denote by $(\cdot,\cdot)_M$ the usual inner product on H_M. Now define $C_M = \{f \in V_M \mid f \geq g \text{ on } (-M,M) \text{ and } f = g \text{ if } x = \pm M\}$. We know that ν_M satisfies the following

$$\nu_M \in L^2([0,T], V_M)$$

$$\frac{\partial \nu_M}{\partial t} \in L^2([0,T], H_M)$$

$$\nu_M(t,.) \in C_M$$

$$-\left(\frac{\partial \nu_M}{\partial t}, w - \nu_M\right)_M + a_M(\nu_M, w-\nu_M) \geq 0$$

for all $w \in C_M$

$$\nu_M(T,.) = \psi.$$

For $h > 0$ let

$$R_h = \{m \in \mathcal{Z} \mid -M \leq (m-\frac{1}{2})h < (m+\frac{1}{2})h \leq M\},$$

$$m_1 = \min(R_h),$$

$$m_2 = \max(R_h).$$

Denote by $\chi_m^{(h)}$ the indicator function of the interval $((m-\frac{1}{2})h, (m+\frac{1}{2})h]$, and let H_h be the vector space spanned by the functions $\chi_m^{(h)}$ with $m \in R_h$. On H_h, the finite difference approximation of the operator

$$A = (\frac{\sigma^2}{2})\frac{\partial^2}{\partial x^2} + (r - \sigma^2/2)\frac{\partial}{\partial x} - r$$

is the operator A_h defined as follows: if $u = \sum_{m=m_1}^{m_2} u_m \chi_m^{(h)}$, then

$$A_h u = \sum_{m=m_1+1}^{m_2-1} (A_h u)_m \chi_m^{(h)},$$

with

$$(A_h u)_m = \frac{\sigma^2}{2h^2}(u_{m+1} - 2u_m + u_{m-1}) + \frac{r - \sigma^2/2}{2h}(u_{m+1} - u_{m-1}) - ru_m$$

for $m_1 \leq m \leq m_2$. The function ψ will be approximated by

$$\psi_h = \sum_{m \in R_h} \psi(mh)\chi_m^{(h)}.$$

Now let k be a positive number of the form $k = T/N$, with $N \in \mathcal{N}$. The finite difference approximation of ν_M is the function $\nu^{h,k}(t, x)$ defined by

$$\nu^{h,k}(t, x) = \sum_{i=1}^{N} \nu_h^i(x)\mathbf{1}_{((i-1)k, ik]}(t),$$

where $\nu_h^1, \nu_h^2, \ldots, \nu_h^N$ are in H_h and satisfy the following implicit scheme:

$$\nu_h^N = \psi_h$$

$$\nu_h^i(m_1 h) = \psi(m_1 h),$$

$$\nu_h^i(m_2 h) = \psi(m_2 h)$$

and for all $x \in ((m_1 + \frac{1}{2})h, (m_2 - \frac{1}{2})h)$

$$\max((\nu_h^{i+1}(x) - \nu_h^i(x))/k + A_h\nu^i(x), \psi_h(x) - \nu_h^i(x)) = 0$$

for $i = 1, \ldots, N-1$.

The above system can be solved using backward regression and the Brennan-Schwartz algorithm. We know that as $h, k \to 0$, the functions $\nu^{h,k}$ converge strongly to ν_M in L^2 (Jaillet et al [26]).

By Proposition 11, for h small enough and all $t \in (0, T-k)$, there exists a number $\rho^{h,k}(t) \in [(m_1 + \frac{1}{2})h, (m_2 - \frac{1}{2})h)$ such that

$$\nu^{h,k}(t,x) = \psi_h(x)$$

for all $x \in (m_1 h, \rho^{h,k}(t))$, and

$$\nu^{h,k}(t,x) > \psi_h(x)$$

for all $x \in (\rho^{h,k}(t), (m_2 - \frac{1}{2})h)$. Since $\nu^{h,k}(t,x)$ is a nonincreasing function of t, $\rho^{h,k}$ is nonincreasing. From the proof of Proposition 11 one also knows that

$$(t,x) \in (0, T-k] \times ((m_1 + \frac{1}{2})h, (m_2 - \frac{1}{2})h) \Rightarrow \nu^{h,k}(t,x) > 0.$$

Therefore $\rho^{h,k}(t) \leq \log(X) + h$. Now define the approximate critical price

$$s^{h,k}(t) = e^{\rho^{h,k}(t)}.$$

Theorem 15 *As h and k go to zero, $s^{h,k}$ converges uniformly to s_M.*

PROOF. Due to the properties of $\rho^{h,k}$ and s_M, it suffices to show pointwise convergence on $(0,T)$. Let $t_0 \in (0,T)$. Assume that $x < \limsup s^{h,k}(t_0)$. Then, for a suitable subsequence, we have

$$\nu^{h,k}(t_0, \log(x)) = \psi_h(\log(x)).$$

Therefore, for all $(t, y) \in [t_0, T] \times (-M, \log(x)]$ one has $\nu^{h,k}(t_0, y) = \psi_h(y)$. Hence, letting $h, k \to 0$, one obtains $\nu_M(t, y) = \psi(y)$ almost everywhere on $[t_0, T] \times (-M, \log(x)]$, and so $x \leq s_M(t_0)$.

It remains to prove that $\liminf s^{h,k}(t_0) \geq s_M(t_0)$. Suppose not, and take $\varepsilon > 0$ such that $\liminf s^{h,k}(t_0) < s_M(t_0) - \varepsilon$. Since s_M is continuous, there exists $\eta > 0$ such that for all $t \in [t_0 - \eta, t_0 + \eta]$ we have

$$\liminf s^{h,k}(t_0) < s_M(t) - \varepsilon.$$

Passing to a subsequence, we may assume that

$$s^{h,k}(t_0) \leq s_M(t_0 - \eta) - \varepsilon.$$

Now if $t_0 - \eta < t < t_0$ and $\log(s_M(t_0 - \eta) - \varepsilon) < x < \log(s_M(t_0 - \eta))$, then

$$x > \log(s^{h,k}(t_0)) \geq \log(s^{h,k}(t)),$$

whence $\nu^{h,k}(t, x) > \psi_h(x)$ and, by the definition of the $\nu^{h,k}$

$$\frac{\nu^{h,k}(t+k, x) - \nu^{h,k}(t, x)}{k} + A_h \nu^{h,k}(t, x) = 0.$$

Now let

$$U = (t_0 - \eta, t_0) \times (\log(s_M(t_0 - \eta) - \varepsilon), \log(s_M(t_0 - \eta)))$$

and let ϕ be a C^∞ function with support in U. It is easy to check (using the fact that $\nu^{h,k} \to \nu_M$ in L^2) that

$$\lim_{h,k \to 0} \int \left(\frac{\nu^{h,k}(t+k, x) - \nu^{h,k}(t, x)}{k} + A_h \nu^{h,k}(t, x)\right)\phi(t, x) dt dx$$

$$= \int \nu_M\left(-\frac{\partial \phi}{\partial t} + \frac{\sigma^2}{2}\frac{\partial \phi^2}{\partial x^2} - (r - \frac{\sigma^2}{2})\frac{\partial \phi}{\partial x} - r\phi\right) dt dx.$$

Hence

$$\frac{\partial \nu_M}{\partial t} + \frac{\sigma^2}{2}\frac{\partial \nu_M^2}{\partial x^2} - (r - \frac{\sigma^2}{2})\frac{\partial \nu_M}{\partial x} - r\nu_M = 0$$

in the sense of distributions on U. But on this set, $\nu_M(t,x) = \psi(x) = X - e^x$, since $s_M(t) \geq s_M(t_0 - \eta)$ if $t \geq t_0 - \eta$. Hence

$$\frac{\partial \nu_M}{\partial t} + \frac{\sigma^2}{2}\frac{\partial \nu_M^2}{\partial x^2} - (r - \frac{\sigma^2}{2})\frac{\partial \nu_M}{\partial x} - r\nu_M = -rX$$

and we have reached a contradiction. □

4.4 Allegretto, Barone-Adesi and Elliott

As usual, $e(S,t)$ denotes the early exercise premium. From work of Carr, Jarrow, and Myneni [15], it is known that $e(S,t)$ has the following integral representation

$$e(S,t) = \int_t^T (rXe^{-r(s-t)}\Phi(-h_2) - (r-b)Se^{-(r-b)(s-t)}\Phi(-h_1))ds, \quad (4.25)$$

where $h_1 = (\log(S/S_c) + (b + \frac{\sigma^2}{2})(s-t))/\sigma\sqrt{s-t}$ and $h_2 = h_1 - \sigma\sqrt{s-t}$. Further, the boundary conditions imply

$$p(S_c,t) + e(S_c,t) = X - S_c, \quad (4.26)$$

$$\frac{\partial p}{\partial S} + \frac{\partial e}{\partial S} = -1 \text{ when } S = S_c\ . \quad (4.27)$$

Begin by trying a function of the form $A(t)(S/S_c)^{q(t)} = e'(S,t)$, say, for e, where $A(t)$ and $q(t)$ are functions to be determined. Substitution of e' in (4.26), (4.27), together with (3.52) leads to the following equation for q:

$$\frac{\sigma^2}{2}q(q-1) - r + bq + (\frac{dA/dt}{A} - \frac{q}{S_c}\frac{dS_c}{dt}) + \frac{dq}{dt}\log(S/S_c) = 0. \quad (4.28)$$

This equation implies that q is not independent of S, so $e(S,t)$ is not of the same form as e'. One may, however, regard e' as an approximation to e. Further, a useful approximation can be obtained by neglecting the last term of (4.28). That is, consider

$$\frac{\sigma^2}{2}q(q-1) - r + bq + (\frac{dA/dt}{A} - \frac{q}{S_c}\frac{dS_c}{dt}) \approx 0. \tag{4.29}$$

This approximation is reasonable when $\log(S/S_c)(dq/dt)$ is small, which happens when either (dq/dt) is small (at long maturities), or in a neighbourhood of S_c. Further manipulation of the equations eventually leads to

$$q^2 + (N-1)q - (M+G) \approx 0, \tag{4.30}$$

where $M = 2r/\sigma^2$, $N = 2b/\sigma^2$, and $G(t) = 2\sigma^{-2}A\partial p/\partial t$. As usual, the boundary conditions at infinity rule out one of the roots, leaving us with

$$q(t) \approx (1 - N - \sqrt{(1-N)^2 + 4(M + G(t))})/2. \tag{4.31}$$

A numerical procedure for finding S_c now suggests itself. Consider the equations

$$A(t) \approx -p(S_c, t) - S_c + X, \tag{4.32}$$

$$q^2 + (N-1)q - (M+G) \approx 0, \tag{4.33}$$

$$S_c \approx \frac{q(t)(X - p(S_c, t))}{-1 + q(t) + e^{(b-r)(T-t)}\Phi(-d_1(S_c, t))}. \tag{4.34}$$

For a fixed value of t, begin with an initial guess for S_c, and successively find new estimates for A, q, and S_c from the above equations. The cycle is to be repeated until the total difference between the successive approximations is less than some pre-assigned value (e.g., 10^{-4}), at which time one has hopefully obtained a good approximation to S_c.

As for the accuracy of the estimates, the integral expression (4.25) for e was evaluated by Bode's rule ([30], pp 649-652), and $(e(S_c, t) - A(t))/A(t)$ calculated. This percentage error was typically of the order of 2%.

To improve the accuracy, the equation (4.30) was modified by the introduction of a relaxation constant λ to yield

$$q^2 + (N-1)q - (M + \lambda G) = 0. \tag{4.35}$$

The same iterative scheme as above was followed, but now for different values of $\lambda \in [0, 2]$, chosen so that $e(S_c, t) = A(t)$.

For example, with $\sigma \in [0.2, 0.4]$ and $r \in [0.04, 0.20]$, various ratios b/r were considered. For $r = b$, the following empirical formula was determined $\lambda = 1.2952 + 4.3338 \times 10^{-2} M - 4.6591 \times 10^{-3} M^2 + 2.1452 \times 10^{-4} M^3$, (recall that $M = 2r/\sigma^2$), and the error $(e(S_c, t) - A(t))/A(t)$ was well below 1% (except for very short times, where problems of computational accuracy arose). For $r = b/2$, and $\lambda = 1.227 + 0.12066M - 4.2737 \times 10^{-2} M^2 + 5.453 \times 10^{-3} M^3$, good results were again obtained. Finally, for $b = 0$, the expression $\lambda = 1.2495 - 4.15 \times 10^{-2} \sigma$ gave satisfaction. Generally, the error was of the order of 0.1% for periods longer than two years. The authors conjecture that the best value of λ is the smallest parameter value for which S_c is monotone.

4.5 Post 1994 papers

Since the first version, there has been several additional work done on the American put. This section will provide a sample of these works.

Davis and Zariphopoulou [14] consider the American options with transaction costs. According to the authors, the perfect replication is no longer possible and holding an option involves an essential element of risk when transaction costs are included. Davis and Zariphopoulou [14] define option price by computing the maximum price at which a utility maximizing investor would include the option in his porfolio. By solving a combination of stochastic control problems with singular policies and optimal stopping, one can compute the option price.

Amin and Bodurtha [2] develop an arbitrage-free discrete time model to price American-style with stochastic interest rates. This general model is *path dependent* and can be implemented with arbitrary volatility functions to value claims with maturity up to five years. Examples of long-dated American currency and warrants and cross-rate swap from the quanto class are included.

Iwaki, Kijima and Yoshida [28] compute the American put options with a finite-set of exercisable time epochs. Dewynne and Wilmott [15] discuss American options with varying exercise prices.

The method of lines (MOL) is a numerical method which applies to initial value (time-dependent) problems. It reduces a partial differential equation to a system of ordinary differential equations. The resulting system of ordinary differential equations is then solved using a computer program (see Rice [59], pp. 524-527 for further discussion on this method). Goldenberg and Schmidt [22], Carr and Faguet [8] and Meyer and van der Hoek [55] discuss how the method of lines can provide a numerical valuation of the American put options and determine the early exercise boundary. Meyer and van der Hoek [55] claim that MOL is flexible, accurate and computationally fast.

Huang, Subrahmanyam and Yu [27] alternatively discuss a recursive integration method to price and hedge American options. They suggest that this recursive integration permits the direct implementation of analytical formulae to compute option values as well as option hedge parameters.

Notation

- t = current time and $t \in [0, T]$.
- T = expiration date.
- $T - t$ = the time to expiration.
- r = risk-free interest rate.
- μ= expected return of the stock price.
- σ= volatility (standard deviation) of the stock price.
- X = exercise (strike) price.
- S_t = stock price at time t.
- S_c = the critical stock price = the price at which the put should be exercised.
- P_t = price of the American put at time t.
- C_t = price of the American call at time t.
- p_t = price of the European put at time t.
- c_t = price of European call at time t.
- $M = 2r/\sigma^2$.
- $f^+ = \max(f, 0)$
- $g(x) = (X - x)^+$.

Bibliography

[1] Allegretto, W, Barone-Adesi, G and Elliott, R J (1994) The free boundary of the American put option. *Preprint.*

[2] Amin, K I and Bodurtha, J N Jr (1995) Discrete-time valuation of American options with stochastic interest rates. *The Review of Financial Studies, 8* (1), 193-234.

[3] Bachelier, L (1900) *Theory of Speculation* English translation published in Cootner [10], 17-78.

[4] Barone-Adesi, G and Whaley, R E (1987) Efficient analytic approximations of American option values. *Journal of Finance, 42*, 301-320.

[5] Bensoussan, A (1984) On the theory of option pricing. *Acta Applicandae Mathematicae, 2*, 139-158.

[6] Black, F and Scholes, M (1973) The pricing of Options and Corporate Liabilities. *Journal of Political Economy, 81*, 637-659.

[7] Brennan, M and Schwartz, E (1977) The valuation of American put options. *Journal of Finance, 32*, 449-462.

[8] Carr, P and Faguet, D (1994) Fast accurate valuation of American options. *Preprint.*

[9] Carr, P, Jarrow, R and Myneni, R (1990) Alternative characteristics of American put options. *Mathematical Finance, 2* (2), 87-106.

[10] Cootner, P H (ed) (1964) *The Random Character of Stock Market Prices.* MIT Press, Cambridge.

[11] Cox, J and Rubinstein, M (1984) *Options Markets.* Prentice Hall, Englewood Cliffs, N. J.

[12] Cox, J, Ross, S, and Rubinstein, M (1979) Option pricing: a simplified approach. *Journal of Financial Economics, 7*, 229-263.

[13] Dahlquist, G and Bjorck, A (1974) *Numerical Methods.* Prentice Hall, Englewood Cliffs, N. J.

[14] Davis, M H A and Zariphopoulou, T (1995) American options and transaction fees. *Mathematical Finance.* Springer, NY. In the *IMA volumes in Mathematics and its applications, Volume 65*, edited by MHA Davis, W H Fleming, D Duffie, and S E Shreve, 47-61.

[15] Dewynne, J N and Willmott, P (1995) A note on American option with varying exercise price. *Journal of Australian Mathematical Society, Series B - Applied Mathematics, 37* (Part 1), 45-57.

[16] El Karoui, N (1981) *Les Aspects Probabilistes du Contrôle Stochastique.* Lecture Notes in Mathematics, 876, Springer, Berlin.

[17] El Karoui, N and Karatzas, I (1989) A new approach to the Skorohod problem and its applications. *Stochastics and Stochastics Reports, 34*, 57-82.

[18] Fama, E F (1970) Efficient capital markets: A review of theory and empirical work. *Journal of Finance, 25* (2), 383-417.

[19] Friedman, A (1975) *Stochastic Differential Equations and Applications I.* Academic Press, New York.

[20] Geske, R and Johnson, H E (1984) The American put option valued analytically. *The Journal of Finance, 34* (5), 1511-1524.

[21] Geske, R and Shastri, K (1985) Valuation by approximation: a comparison of alternative option valuation techniques. *Journal of Financial and Quantitative Analysis, 20* (1), 45-71.

[22] Goldenberg, D H and Schmidt, R J (1994) Estimating the early exercise boundary and pricing American options. *Preprint.*

[23] Gnedenko, B V and Kolmogorov, A N (1954) Limit distributions for sums of independent random variables. Translated by K L Chung. Addison-Wesley, Reading, MA.

[24] Hamming, R W (1973) *Numerical Methods for Scientists and Engineers* (2nd edition). McGraw-Hall, N.Y.

[25] Harrison, J M and Pliska, S R (1981) Martingales and stochastic integrals in the theory of continuous trading. *Stochastic Processes and their Applications, 11*, 215-260.

[26] Hull, J (1993) *Options, Futures and Other Derivative Securities* (2nd Edition). Prentice-Hall, Englewood Cliffs, N.J.

[27] Huang, J, Subrahmanyam, M G and Yu, G G (1996) Pricing and hedging American options: A recursive integration method. *The Review of Financial Studies, 9* (1), 277-300.

[28] Iwaki, H, Kijima, M and Yoshida, T (1995) American put options with a finite-set of exercisable time epochs. *Mathematical and Computer Modelling, 22* (10-12), 89-97.

[29] Jacka, S D (1991) Optimal stopping and the American put. *Mathematical Finance, 1* (2), 1-14.

[30] Jacka, S D (1993) Local times, optimal stopping, and semimartingales. *The Annals of Probability, 21* (1), 329-339.

[31] Jacka, S D and Lynn J R (1992) Finite-horizon optimal stopping, obstacle problems and the shape of the continuation region. *Stochastics and Stochastics Reports, 39*, 25-42.

[32] Jaillet, P, Lamberton, D, and Lapeyre, B (1988) Analyse numérique des options américaines. *Cahier du CERMA No. 9*, 66-126.

[33] Jaillet, P, Lamberton, D, and Lapeyre, B (1990) Variational inequalities and the pricing of American options. *Acta Applicandae Mathematicae, 21*, 263-289.

[34] Johnson, H E (1983) An analytic approximation for the American put price. *Journal of Financial and Quantitative Analysis, 18* (1), 141-148.

[35] Karatzas, I (1988) On the pricing of American options. *Applied Mathematics and Optimization, 17*, 37-60.

[36] Karatzas, I (1989) Optimization problems in the theory of continuous trading. *SIAM Journl on Control and Optimization, 27*, 1221-1259.

[37] Karatzas, I (1990) Applications of stochastic calculus in financial economics. In *Recent Advances in Stochastic Calculus*, edited by J S Baras and V Mirelli, 169 -217.

[38] Karatzas, I (1996) *Lectures on the Mathematics of Finance.* In the CRM Monograph Series, Volume 8. AMS, Boston, MA.

[39] Karatzas, I and Shreve, S E (1994) *Monographs on Mathematical Finance.* In preparation.

[40] Kim, I J (1990) The analytic valuation of American options. *Review of Financial Studies, 3*, 547-572.

[41] Kim, I J and Yu, G G (1993) A simplified approach to the valuation of American options and applications. *Preprint.*

[42] Kinderlehrer, D and Stampacchia, G (1980) *An Introduction to Variational Inequalities and their Applications.* Academic Press, New York.

[43] Krylov, N V (1980) *Controlled Diffusion Processes.* Springer, New York.

[44] Kushner, H J (1977) *Probability Methods for Approximations in Stochastic Control and for Elliptic Equations.* Academic Press, New York.

[45] Lamberton, D (1993) Convergence of the critical price in the approximation of American options. *Mathematical Finance, 3*(2), 179-190.

[46] MacMillan, L W (1986) Analytic approximation for the American put option. *Advances in Futures and Options Research, 1* (A), 119-139.

[47] Margrabe, W (1978) The value of an option to exchange one asset for another. *Journal of Finance, 33*, 177-186.

[48] Mason, S (1978) The numerical analysis of certain free boundary problems arising in financial economics. *Working paper 78-52, Harvard Business School.*

[49] McKean, H P, Jr (1965) Appendix: a free boundary problem for the heat equation arising from a problem in mathematical economics. *Industrial Management Review, 6* (2), 32-39.

[50] Merton, R C (1969) Lifetime Portfolio selection under uncertainty: the continuous-time case. *Review of Economics and Statistics, 51* (3), 247-257.

[51] Merton, R C (1971) Optimum consumption and portfolio rules in a continuous-time case. *Journal of Economic Theory, 3* (4), 373-413.

[52] Merton, R C (1973) Theory of rational option pricing. *Bell Journal of Economics and Management, 4*, 141-183.

[53] Merton, R C (1977) On the pricing of contingent claims and the Modigliani-Miller Theorem. *Journal of Financial Economics, 5*, 241-249.

[54] Merton, R C (1992) *Continuous-time finance* (revised edition) Blackwells Publishing, Cambridge, MA.

[55] Meyer, G H and van der Hoek, J (1996) The evaluation of American options with the methods of lines. *Advances in Futures and Options Research* (to appear).

[56] Miller, M H and Modigliani, F (1961) Dividend policy, growth, and the valuation of shares. *Journal of Business, 34*, 411-433.

[57] Myneni, R (1992) The pricing of the American option. *The Annals of Applied Probability, 2*, 1-23.

[58] Parkinson, M (1977) Option pricing: the American put. *Journal of Business, 50*, 21-36.

[59] Rice, J R (1993) *Numerical Methods, Software, and Analysis* (2nd Edition). Academic Press, San Diego, CA.

[60] Protter, P (1990) *Stochastic Integration and Differential Equations: A New Approach.* In the Applications of Mathematics, Number 21, Springer-Verlag, New York.

[61] Samuelson, P A (1965) Rational theory of warrant pricing. *Industrial Mangement Review, 6* (2), 13-31.

[62] Samuelson, P A (1965) Proof that properly anticipated prices fluctuate randomly. *Industrial Management Review, 6* (2) 41-50.

[63] Samuelson, P A and Merton, R C (1969) A complete model of warrant pricing that mazimizes utility. *Industrial Management Review, 10* (2), 17-46.

[64] Sharpe, W F (1985) *Investments* (3rd edition). Prentice Hall, Englewood Cliff, NJ.

[65] Smith, C W (1976) Option pricing: a review. *Journal of Financial Economics, 3*, 3-51.

[66] Van Moerbeke, P (1976) On optimal stopping and free boundary problems. *Archives Rational Mechanics and Analysis, 60*, 101-148.

Index